The Wisdom of Loki

The Art of Lying in the Natural World

Ptera Hunter, PhD

The Wisdom of Loki

The Art of Lying in the Natural World

Ptera Hunter, PhD

Benjamin Ross & Lane Publishers

Cover art by Ptera Hunter

Edition: Jan 2023
ISBN: 9780-9629-5555-6

Published by Benjamin, Ross & Lane
www.benjaminrosslane.com

Acknowledgements

In alphabetical order, I thank Mr. J.A. Cebuhar and Mr. K.J.J.J. Kelling for reviewing this paper from the perspective of a layman. I further thank Dr. R.I.C. Hansel of Noble International University, and Dr. E. McArdle, emeritus Northeastern Illinois University for reviewing the science. I also acknowledge the contribution of Ms. Angela Pettera for her invaluable aid in editing this book and for reminding me that there is more than one word to express most thoughts.

Dedication

Dedicated to Dr. Eugene McArdle, a good friend, a trusted advisor, and someone who helped me through some of the toughest times in my life. Thank you for having confidence in me, even when I had none in myself.

And to the Wisdom of Loki.

"The power of beauty will sooner transform honesty from what it is to a bawd than the force of honesty can translate beauty into his likeness: this was sometime a paradox, but now the time gives it proof."

~ *Hamlet,* Act 3, Scene 1, by William Shakespeare

"In the desert, an old monk had once advised a traveller, the voices of God and the Devil are scarcely distinguishable."

~ *The Unexpected Universe: A Library of America eBook Classic*

by L. Eiseley

Contents

Preface

This book describes one of the most basic activities of all living things: lying. The art of deception runs through the world of living things from the most complex of mammals and birds through bacteria and DNA itself.

Cole Porter's 1928 song "*Let's Do It*" kept going through my mind as I worked on this the book. Mr. Porter meant the song to describe love and sex. It also serves as an allegory for how the living world uses miscommunication to its benefit for survival and reproduction. I could tell you that I chose this song because the song serves as an analogy for the deep evolutionary roots for lying.

I could also tell you that the idea struck me as whimsical, or that it sounded like a fun way to organize my thoughts. If I were to make such an admission, I would be illustrating a seminal point in the book: When individuals choose an honest path, the outcome is not always the best. If an organism presents itself honestly, complete with shortcomings, it becomes an open target. The competition can find the soft underbelly and knows where to strike.

If I admit a whimsical approach to the organization of this book, then fewer readers would take the content—or me—seriously. Instead, by stretching the truth and by appealing to deep intellectual roots, I make my idea sound important. The wording intimidates people, so they see me as a brilliant, albeit pompous, theorist who understands the nature of humankind and the universe. From my perspective, that outcome is far more desirable.

So which path did I take? I took the brilliant, pompous approach, with its recognition of the deep evolutionary psychological roots of Cole

Porter's music—of course. As the old Le Roy Parnell song puts it, "that's my story and I'm sticking to it" (L. R. Parnell, released 1993 by Collin Raye).

Across the world of living things, much communication consists of partial truths or untruths and efforts to detect them (1; 2). Because we can learn so much more than most other animals, vertebrates are very gifted at deception. We use our brains not just to communicate, but also to miscommunicate information. As vertebrates, we lie about our status, our professional ambitions, and our sexual intentions. We lie about our gender. We even lie about our ability to defend our homes and our families.

There is an old saying that everyone hates a liar. If this adage were true, then we must hate ourselves. For virtually every species, lying is an essential survival skill. Even yellow-jacket wasps will attack a low-ranking wasp that sports the colouration of a higher-ranking wasp (3). In the "*Pirates of Penzance,*" writers Gilbert and Sullivan (4), poke fun at General Stanley—the Modern Major General who is wracked with guilt night after night because he had to tell a lie. That lie prevented pirates from ravaging his daughters. Gilbert and Sullivan clearly conveyed the message that although everyone recognizes lying as a social wrong, no one in his or her right mind would have preferred the alternative outcome of rape and forced marriage.

In our society, lying and truth-telling find a balance. We know we should not lie under oath, but the questions asked in such a situation should also be limited to areas that at least seem to have relevance. We cast a jaundiced eye on members of the legal profession who abuse the power of their office to ruin a life or a reputation. People often see these offenders in a more cynical light than the people whom they interrogate.

People are not the only liars on the planet. Plants, fungi, and even unicellular beings also practice deceitful behaviours. This book will use Mitchell's definition of deception as any false communication that benefits the communicator, even if the transmission of false information is not intentional (5).

This book will present a brief and very incomplete overview of lying, deceit, and deception in the natural world. It will focus on how species use deception to gain an advantage. The book aims at presenting an overview of the works on trickery produced by giants in this field of study. The purpose is not to make us better liars, but to help us understand the underlying evolutionary pressures that lead to deceptions across the natural world.

We can never eliminate ruses, nor would their total elimination necessarily be desirable. Instead, this book will try to explain them. Once we comprehend the evolutionary function of deception, we stand better equipped to understand why it happens, when it is likely to happen, and who is likely to deceive us. The more we appreciate the nature of deceit, the better prepared we are to deal with its consequences on our lives. It gives us the tools we need to confront it or at least to defend ourselves against it.

A Note on Lying

Lying includes all of the things we do or say to mislead others, no matter what the reason. It also encompasses those things we do not say or do for the same reasons. Sometimes we lie to help others, and sometimes we lie to help ourselves. No matter what we call it, lying occurs when we mislead other beings.

Many synonyms make us feel better about lying. Below, I list a *few* of these synonyms. This book will use them interchangeably to achieve the nuances that we perceive in deceptions.

Lying includes deception, falsehoods, dishonesty, prevarication, distortion, forgery, misrepresentation, deceit, and telling falsehoods, fictions, untruths, crammers, inveracities, tall tales, fibs, confabulations, and white lies.

It includes warping the truth and putting on a sham or a con or being bogus or fake.

It means engaging in pretence, make-believe, ruses, simulations, imitations, mimicry, impersonation, trickery, hoaxes, duplicity, counterfeiting, swindling, and snolygostering. (Yes, that last one is a really is a word, or at least a gerund formed from the word for a corrupt, lying politician.)

Lying comprises being phoney, falsifying documents, copying other's work, plagiarism, and twisting or slanting or spinning or bending the truth.

It also includes frauds, spurious or misleading ideas, Munchausen's disease, inflated bragging, fairy stories, flim-flam and just plan flam.

It includes artificiality, half-truths, intentional oversights, and lies of omission, to name just a few.

Chapter 1

Lying and the Origins of Deceit

"When the little blue bird
Who has never said a word
Starts to sing"
~ "Let's Do It" by Cole Porter

Oden, Loki, and Thor: A Parable

After the great tree of the universe had blossomed with life, Oden began his journeys among the nine realms. He traveled across its branches to learn about the kingdoms of the Gods, the lands of men, and the worlds of the demons that devour. Through self-sacrifice, he gained the power of written knowledge from the well of Urd from which all life sprang. The well rewarded his effort and commitment, and the runes found him worthy. He then knew what had passed before and could contemplate it.

Because he had knowledge, Oden also craved its sister, wisdom. He returned to Urd. He asked Mimir, the guardian of the well, for a drink of wisdom. Mimir refused because he understood the power that wisdom could unleash should it fall into the wrong hands. Only if Oden were to gouge out an eye would the guardian fulfill his desire. For Oden, no sacrifice was too dreadful if it bestowed wisdom. Oden knew he would still have one eye to see the present. Therefore, he complied, and then he drank. Thus, he then gained wisdom and, with it, forethought.

Through his knowledge and wisdom, Oden proved himself worthy to rule in Asgard. These qualities showed him that to retain the Asgardian throne, he must spread discord among his enemies and bring the Gods together.

Wisdom also taught Oden that he could not rule alone. Even the wisest chieftain needs allies. He ruled with the consent of all the Gods, but his closest allies were Thor and Loki. Both were born of the realm of chaos, both from mothers who were of the giants who devour.[1] These two aids would bring to Asgard the other skills needed to rule.

Thor sprang from the seed of Oden and had a share of his wisdom. Thor had strength and respect. Although born of the demon, he lived as a son of Asgard. He protected the weak and could vanquish even the mightiest giant. His knowledge of order helped him as he led the Gods into their battles. Oden could rule, but Asgard depended upon Thor, the warrior, for its stability, organization, and strength.

The third member of the triumvirate was Loki. In battle, Oden had killed Loki's father, Fárbauti the Cruel. Oden then brought Loki to Asgard to live among the Gods. Loki, the trickster, became the blood-brother of Thor, and Thor pledged to protect him. Many times, Oden had to remind Thor of that promise.

Loki had a mind that worked as fast as fire. He listened carefully, and he could use the words of others against them when he needed to do so. He entrapped them as skillfully as a spider nets its prey. Loki did not

[1] Sources are a bit iffy on this point. Some traditions hold that Laufey, the mother of Loki, was a goddess. Others, that she was a giant.

suffer from morality. He shifted his shape and his allegiances as needed. Loki had little loyalty or honour.

The Gods of Asgard wondered why Oden protected Loki. They could not understand why the noble Thor held Loki as his beloved blood brother. Many times, Loki created havoc throughout Asgard and endangered the Gods. They knew Loki as the entangler and as the King of Lies. Because they lived without the wisdom possessed by Oden alone, the other Gods could not understand the purpose of Loki nor his place by Oden's side.

Despite his amorality, despite his shameful character, Loki often saved Asgard from destruction. Loki alone mastered the skills of trickery and deceit. Oden might know the tactics of demons and men, but only Loki could turn their plans against them. Thor knew how to organize an army; Loki knew how to keep their tactics a secret and how to mislead their enemies. Without the ability to deceive, the foresight of Oden had no power to protect Asgard. Without the ability to lie, the power of Thor stood greatly diminished. Loki had a distinct and specialized knowledge and a wisdom that even Oden lacked. Loki had the knowledge of deceit, and he had the wisdom of how to apply it to gain and keep his advantage.

The Gods could not trust the King of Lies. Eventually, his deceit and lack of honour would bring about the downfall of the kingdom of the Gods. However, without his aptitude for deception, they would have lost Asgard many times. Thor's hammer gained strength to defend the Gods through the ability to cheat and mislead that only Loki could provide. Loki understood how to use the foresight of Oden better than did Oden. Loki could use that knowledge to understand his quarry and to gain an advantage. His wisdom lay in how to combine the skills of Oden and Thor with his talent for trickery.

I dedicate this book to my friend, Eugene McArdle and to the Wisdom of Loki.

The Pre-Verbal Origins of the Lie

For years, people believed that the power of communication lay in its ability to transfer accurate information from one individual to the next. That answer seems obvious. On its return from its "hunt," the honey bee gives its hive-mates the location of the juicy flowers by doing a dance that informs them of the location of the nectar-laden treats. We attempt to keep our children safe by communicating danger to them. Birds raise the alarm to transmit information about a threat to the other members of its flock. What could natural selection favour more than the honest transmittal of information?

In each of these examples, the interest of the communicator mirrors the interest of the animals receiving the message. The scout bee and the hive mates all want to gather that honey. The parent wants to keep its children safe, and like all humans, children want to maximize pleasure while avoiding harm. The birds in the flock want to avoid danger, and the communicator hopes that the hoard of panicked flying birds will make it harder for a predator to home in on it.

When the desires and the needs of the message-sender and the message-receiver coincide, then honest communication works well, and honesty is the preferred mode of communication. If a disconnect exists, then the sender can derive a benefit from sending disinformation to the receiver. Before I can even hope to lie, I need to be able to communicate.

Before I can expect to understand a lie, I need to understand the intent of my neighbour. To achieve these ends, animals communicate with systems that range from muscle movements and scents to chemical signals as old as life itself.

Evolution grafted our communication skills onto previously existing systems. As animals, we breathed. As animals, we also learned to interpret the breathing of other animals. From this basis, we derive information about those around us. If our neighbour gasps for breath, it probably means he or she just stopped running. Could he have been evading a predator? Could she have been running down prey? Or did he just finish playing a game? By looking at all the signals, we can make interpretations.

We also developed an ability to use our breathing to send signals. If we alter breathing one way, we imply aggression. If we change it differently, we show pleasure. We also learned to use other bodily systems in the same way. Urinate on a tree, and it will mark territory. Adopt a physically defensive posture and signal submission. Even in the absence of language, animals use these physiological and anatomical traits to express themselves.

Organisms that could modify an already existing physiological system, such as vision or smell, into a signal or sound, had an advantage over those organisms that had to learn how to communicate from scratch. Once in place, animals quickly adapted to these abilities and used these mechanisms to trick others into believing what is not true.

One Small Step: Birth of the Verbal Lie

Miscommunication has an extreme adaptive advantage. Signals can misdirect the receiver and cloak the real intent of the signaler. When biologists first studied the origins of communication, they looked at those specific points where interests overlapped. These biased selections of data led them to believe that accurate communication was the norm.

They were wrong. Among animals, conflict over resources and diverging interests are at least as frequent as converging interests. When we think about communication in this light, it becomes far more interesting. Here is where lying comes into its own. Conflicts over resources are so pervasive in the natural world that deceit is not just an option. Deceitfulness and lie-detection are both essential survival skills.

The elk that signals dominance with a long bray needs to fight less often than the elk that does not signal a tough-guy attitude. With a good bluff, he can usually avoid a costly fight. Fiddler crabs that have lost a claw will re-grow a bigger but weaker claw. When challenged by a smaller crab, the weak-clawed fiddler will bluff. It will display a fighting stance that tells the smaller crab that a confrontation would be foolish (6).

As we learn to lie, we also learn to become lie detectors (7). Any animal that cannot tell a bluff from an honest signal is at a considerable disadvantage. Bolo spiders successfully hunt their moth prey because they produce a false mating signal that lures local male moths into their traps (8). In one species of firefly (genus *Photuris*), females can dine on the males of another species for this same reason: they send a phony mating signal to the other species (9-13). The duped male becomes dinner rather than a date.

One katydid (*Chlorobalius leucoviridis*) hunts male cicada with even more skill. This versatile hunter flicks its wings to imitate the mating call of the female. It is such an adept impersonator that it can even replicate the songs of cicada species it has not previously encountered (14). As that male gets closer, this katydid also moves her abdomen with a series of jerks that mime receptive sexual movements of the female cicada.

The chimp that can bluff with a stern face and bristle its hair may be able to intimidate a rival without expending energy or risking a possible injury on a fight (15). In each of these cases, the signaller benefits from duping the receiver of the signal. Both the signaller and the receiver must co-evolve with the other. If the abilities of the receiver lag behind the skills of the liar, he will suffer a disadvantage in life. If the cheater cannot fool the truth-detecting receiver, then his or her hoax will cease to work.

Signal-senders and signal-receivers must thus engage in an arms race (16; 17). As predators become better tricksters, they change the selection pressure on their prey. Prey who can detect the predators, the better lie-detectors, survive to reproduce (16). The reverse is also true. Predators must learn to see through the tricks of the prey. The better the illusion, the more attentive the predator must be. In this way, systems of liars and lie-detectors can push their participants toward more extreme miscommunication. This system can spiral out of control, just like some other animals—such as elephants and Apatosaurus—got stuck on a path toward more extreme size to avoid predation (18), or—like the peacock—got trapped in a cycle of ever greater ornamentation to attract more and more selective mates. Circumstance forced each of these species into their own biological arms race. None of them could end this race for survival without suffering the consequences: extinction.

Some species have developed early warning systems that punish cheaters. Specific signals tell the surrounding population that so-and-so cheats. Some songbirds, for example, can have specific songs that alert the neighbours that someone is violating territorial or sexual boundaries. These songs change the behaviour of the surrounding members of the community, and life becomes rough for the accused cheater (19). As with humans, neighbours know their neighbours. They reject accusations of known intruders, and when in doubt, they tend to take the word of the more dominant bird.

An animal that cannot discriminate between accurate and misleading signals will have problems. The passive acceptance of a message as truth sorely compromises its ability to compete. It will need to treat each threat as viable, and it will assume that each friendly face is indeed friendly. If rival animals know its intent, then that disadvantage becomes magnified because he or she will be less able to gain access to resources than if they could bluff. The resulting cascade of false assumptions and incorrect choices will harm its ability to reproduce and to leave behind its genes in the form of offspring.

An animal that can successfully deceive also has the upper hand over the one who always tells the truth. The female firefly that lures the hapless conspecific male succeeds only in her duplicity about ten percent of the time (20). The trick does not need to work perfectly. Those males that fall for the trick even once become her dinner and calories to support her next generation. The potential targets who recognize the trick and stop short of her survive the challenge of natural selection that she presents. They contribute the genetics that support their superior lie-detection skills to their offspring.

False signals can cloak the real intent of an organism. Lies help us to survive the hunter and to capture our prey. Without the ability to deceive and to detect deceit, we would soon be dead.

Chapter 2

Advantages of Lying in the Natural World

"It is nature that is all
Simply telling us to [lie]"
~ "Let's Do It" by Cole Porter

People who believe that animals cannot lie must not have spent much time with animals. Like humans, animals will lie when it suits their needs. They will lie to avoid challenges or fights and to procure more food or mates. Vertebrates are very gifted at these deceptions. Primates adeptly misrepresent social status, sexual intentions, and ambitions (21; 22). We lie about our gender. We mask our presence. We even exaggerate our ability to defend our homes and our families. Lying is so pervasive that it must serve one or more functions which help organisms survive and thrive in the world of the living.

Because of all the lying, we have learned to defend ourselves from the deceptions of others. (This book will define lying as the miscommunication of information to mislead another living being.) The target of the deception and the modality of delivery do not matter. Like the misdirection of a magician, the point of such communication is to make an organism believe *A* without question when what is really happening is *B*.

We spend an inordinate part of our time trying to determine the reliability of communications by other beings. As vertebrates mature, we learn to recognize the dishonest signals from our neighbours; as we master these cues, we become more adept at lie detection (23; 24). Our large brains allow us to learn. Part of what we learn includes whom we can trust, what is safe, and the outcome of previous errors in judgement.

The early twentieth-century psychologist, Erik Erikson foreshadowed this concept in his model of human development. Erikson, like his mentor Freud, hoped to find a way of understanding behaviour that was less mechanistic than the anatomical approach of his day. Unlike Freud, who saw the tasks of life linked to the movement of pleasure centres across the body, Erikson concluded that our neural development is rooted firmly in our social environment. The psychosocial model of Erickson places the development of appropriate levels of trust at the root of our maturation (25; 26). In this model, what we learn to trust and to mistrust forms the foundation upon which we graft the other tasks of life. Erikson argued that during our infancy, when our brains are the most pliable, we determine how much trust we can place in the world through our early interactions with caregivers. These formative experiences create the personal ecological context that shapes the synapses and the subsequent branching of the neural networks, thus channelling later development (27). For example, if we do not trust our perceptions of ourselves, we are likely to accept shaming or degrading descriptions imposed upon us by others.

Although Erikson saw our ability to trust *appropriately* as our first crucial developmental phase, he never said that this skill developed only in infancy. Each time we meet someone or something new, we re-enact this phase. Using the skill set developed through all our previous experiences, we decide how much trust we can place in the new relationship. As we age, we learn to use subtle but precise visual cues to determine the trustworthiness of another person or another animal. We appreciate that certain eye movements mean danger and others indicate honesty. We come to understand that some signals cross the species barrier, that other signals hold across human cultures, and that others differ among cultures (28).

The Art of Lying

As we learn how to trust, we also learn how to deceive. Jackson (29) argued that schoolchildren learn to mislead others as an essential skill. Children lie to maximize praise and to reduce punishment, to broadcast their successes, and to hide their failures. Through the art of lying, we learn how to please both our peers and our teachers simultaneously. As we learn to understand the signals for trust, we also learn how to fake those signals and how to determine if those false signals have the impact we desire. In other words, we learn to judge the success of our deceptions.

This skill develops, in part, as we learn to take the perspective of another person. In this sense, the learning that makes lying possible must also be present to experience empathy and compassion. Unfortunately, the process does not necessarily work in reverse. The ability to lie may give us the foundations for empathy, but it does not guarantee empathy. A small percentage of the population has a condition called antisocial personality disorder which allows them to understand the perspective of others without feeling empathy for them. The lack of compassion in people with this disorder hallmarks them with callousness and reduced arousal by the distress of other people (30). People with antisocial personality disorder understand what other people think and feel, as evidenced by their ability to manipulate those people. The antisocial personality does not care about those feelings. This disconnect is so striking that some clinicians think about antisocials as "mechanical men" or as emotionally retarded. Although they may understand that an action is wrong, they do not feel it (31; 32). Without this connection between morality and feelings, the antisocial person cannot make the customary use of information about emotions (33).

Brain scans, such as MRIs, show that people with antisocial personality disorder respond differently to language as compared to people close to the behavioural norm. Their brains use words differently. Antisocial personalities also differ in the structure and function of the corpus callosum, an area of the brain that links the right and left sides of the brain (34). This fractured connection prevents them from fully integrating the emotional with the cognitive information of language. This same break can make them masterful liars.

During ordinary communication, the two hemispheres of the brain process words differently. Together, they connect the emotional and the mechanical meanings of the interaction because these differences exist in how these regions work together to interpret one's actions and words (35). Because of abnormal connections in the antisocial brain, the right region that controls our emotional understanding of words does not correctly link to the part that gives words their literal meanings, the left temporal lobe. The areas that control the emotional content of words do not show the same activity in the MRI of an antisocial person as they do in a "normal" person (35). These disconnects in function, anatomy and emotion do not merely make lying easier for the antisocial person. They make it a virtual non-issue.

Functioning psychopaths have some definite advantages. A good con artist can manipulate people into giving him money, food, sex, or any other of the pleasures of life. The scams do not need to work 100 percent of the time. Just as with the hungry deceptive firefly, these human predators do not need to be perfect in their deceptions. A successful approach requires only that their scams work often enough to get the basics for life and reproduction.

We say a life strategy is successful if it leads to the production of a disproportional number of surviving offspring because babies are our genetic ticket to staying in the next generation and the gene game. Successful strategies, either physical or behavioural, spread through a population and supplant less successful tactics. The ability to manipulate others without remorse provides a significant advantage in gaining resources and sex. A smart psychopath will excel in such manipulation.

So why does that particular combination of genes that lead to psychopathy not result in the behaviour spreading through the population, leaving everyone lying to each other, manipulating each other, and just being plain stinkers? Psychopaths do not dominate because they need victims to survive. Their manipulative ploy only works if some dupes also endure in the population. If a population gets too many liars and cheats, then the system that supports the liars becomes unstable. The selection advantage drops like a rock (36). When other people (or other social animals) realize the person's true nature, potential mates usually dislike him.[2] People eventually stop giving aid and comfort in response to his pitiful cries. When the jig is finally up, the reproductive and survival benefits go down. The family and former friends of the antisocial person label him as a cold-hearted liar, and no one trusts him. Group pressures thus restrict the adaptive value of psychopathy and the success of the lying that accompanies it.

Even those of us with standard empathy skills learn and perfect the necessary art of deceit. We learn the difference between types of lies: we distinguish between self-serving lies and pro-social lies. We also learn that these two types of deception have a delicate balance. When we lie for a self-serving or anti-social reason, we feel different than we lie for a pro-social reason that helps someone (37). By adulthood, many of us can work with colleagues we dislike and still see the world through their eyes. The combined skills of understanding and lying help us tolerate those individuals for whom we have no genuine respect without letting them know that we think they are idiots or bullies.

The Universality of Lying

Other species can also learn to lie and to recognize deceit. Social species gain a clear benefit from this skill because they must manoeuvre through the tangled labyrinth of relationships in which they live. Organisms can also derive benefits from accurately interpreting the signals in situations that have little or nothing to do with social

---

[2] Throughout the book, I use "he" and "she" interchangeably rather than using the awkward "he or she" convention at every turn. (Admit it. It gets annoying when you read that. However, in the case of psychopaths, let your biases fall toward "he." Males do form the overwhelming majority of people diagnosed with antisocial disorder. Most of the data about the disorder come from male subjects. We know far less about female psychopaths.

interactions. Gazelles must understand how predators ambush their prey and how they camouflage themselves. Males and females of every species, not just the social species, must decide if signals for mating are accurate indicators or dangerous deceptions.

Rudimentary learning of this kind even occurs in insects. In some species of wasps, the males will repeatedly pollinate a particular tongue orchid. The orchids lure male wasps by mimicking the scent of a female wasp. Some tongue orchids also have patches on their flowers that look like centrefolds of female wasps. The colouring, shape, lines, and texture of the orchid all combine to look like the voluptuous genitalia of a female of his species. The male wasp cannot resist the temptation. He comes by the flower expecting to have sex and then begins mating with it. During this process, the wasp wastes sperm that he might otherwise use to fertilize a real female. Eventually, the wasp does seem to learn that the orchid is not his lady love, and later makes better choices of sexual partners (38; 39). At last count, the family of orchids seem to fool at least 200 species of insects in this way.

One might think that such deception would also be ruinous for the orchid. Instead, the plant benefits by reducing the number of successful copulations by male insects. If a female wasp mates with a male, then half of her offspring will be of each gender. When a female wasp cannot find a male partner, she can still lay unfertilized eggs. All these eggs will develop into males. This process doubles the number of males available to the orchid to spread its pollen. The orchid wins. The female wasp also wins because she has a better chance of passing on her genes by bearing more sons. Males can produce much sperm and mate with many females, but the number of eggs they can lay limits female offspring. The only loser in this situation is the poor duped male who dates the flower.

Chapter 3

The Advantages of Lying and Its Molecular Signalling System

"And that's why birds do it"

~ "Let's Do It" by Cole Porter

The Upside of Lying

The advantages of deceit start at the molecular level. These molecular mechanisms signal us through the cells of the brain, in the secretion of hormones and neurotransmitters, and in all the other chemicals that send messages throughout our bodies. They pave the way toward opportunities for successful avoidance of predation and conflict, and help improve our chance of successful reproduction.

Any organism that can successfully leave babies behind has won round one of the gene game. Round two happens when that baby also reproduces successfully. If that original organism becomes a great-grandparent, it has succeeded at the game of survival.

Living beings navigate the world with a programmed desire to survive and to have sex. Typically, they are not cognizant of this drive to reproduce. Birds have no conscious desire to continue their genetic line when they nest. They merely follow an internal program that tells them to mate, nest, care for their young, and repeat the cycle as necessary. Bees are unaware that they perpetuate their genes by gathering pollen and building combs. None of these organisms mate and have offspring because they knowingly want to perpetuate their genes. It does not matter. The effect remains the same: they act *as if* they were programmed to be carriers of their genes.

The Molecular Basis for Deceit

The molecules of protein that comprise our bodies have no brains and no consciousness. The tadpole that changes colour depends upon its proteins and DNA for its transformation (40), not upon conscious will. Those molecules do not even know that they exist, yet they do have a purpose in their current molecular incarnations. DNA makes a wide range of proteins to regulate the body of its hosts (like us). From generation to generation, we carry this DNA code in our genome and its resulting proteome, the sum of its resulting proteins. These proteins are the stuff that makes the brain, the hormones, the lungs and the heart. Our DNA and its protein products interact with the environment. These interactions influence every aspect of our lives.

Those DNA codes that help their host to survive and to reproduce will also survive the game of life. If the DNA instructs the cell to make a faulty or maladaptive protein, or if it codes for a gene that cannot compete with the genes in other gene-carriers, then that DNA is doomed along with its host. The host will not procreate, and the continuum that has existed for over 4.5 billion years of life will end for that bit of code.

The survivors into the next generation will be the winners, and their genes will be the ones that lead to survival.

Prairie Vole Love

~ with apologies to Willis Alan Ramsey

Sometimes our genes deceive us with bald-faced lies. This penchant is particularly true when it comes to procreation. Our genes can encourage us to believe something is real against our better judgement and our common sense. The molecules in our body encourage us to ignore what we have learned about people and to trust our gut instincts at the expense of reason. For example, males—be they human, animal, or plant—can benefit from having multiple mates. It gives them a chance to have more offspring. Despite the apparent benefit of having multiple females caring for their offspring, there are a few, rare species in which the males habitually practice some form of monogamy.

In mammalian fidelity, molecules, such as vasopressin (41; 42; 43; 44), tell males that being a faithful spouse will help them in their goal to leave offspring behind. High vasopressin activity correlates with that sense of well-being that mammals interpret as "being in love" (45). If you release this hormone into one particular part of the brain, the lateral septum (46), then the male vole—and many other male mammals—will pair-bond with and remain faithful to a female, even at the expense of mating more often or creating more babies.

Female brains respond differently to these internal signals than do male brains [e.g., vertebrates, (47).] Females seem to respond more strongly to another molecule. In pair-bonding species, female brains release oxytocin (45), the hormone associated with the female orgasmic response (48; 49; 41). When the female brain releases oxytocin in the presence of a wooing male, it produces a sense of love and fidelity in the female (50). In humans, oxytocin tricks its female carrier into believing that yes, indeed, he does love her and will remain faithful to her even after having sex.

In females of some species, triggering these molecules can also trigger the release of other molecules in a kind of molecular conspiracy. In shrews, the oxytocin-high can later result in a surge of a releasing hormone called GnRH, or gonatotrophic releasing hormone. GnRH is the chemical that regulates reproduction in virtually all vertebrates and in some invertebrates too. The levels of this hormone change with our ability to reproduce successfully and keeps us from mating when our energy reserves are too low (51). In nutritionally starved female shrews, the GnRH level goes way down. The female then feels less love from a companion shrew. If one then injects those hungry females with GnRH, they switch gears and become sexually receptive (51). The higher GnRH levels also make the female shrews more likely to ovulate (52).

Think about how these molecular signals for love and fidelity interact with the poor impulse-control present in youth. Teenagers are in their prime for reproducing, chock-full of GnRH and therefore filled with feelings of love and fidelity. Suddenly, we can better understand how young bodies trick people into procreative activities and into becoming another generation of DNA-carriers. We can also see why females who diet, starving themselves of food and consequently GnRH, often feel unloved or unlovable.

Molecules do not stop their deceit at initiating or encouraging sex. Sex and procreation through their new hosts will not happen unless they survive to adulthood and through the process of reproduction.

Before they can achieve reproduction, living beings must first thrive to adulthood. To thrive, animals need food. Organisms feel this drive to acquire food on a molecular level. To see this drive in action, one only need be on a university campus at the right time. Just try to manoeuvre against a crowd of students and teachers who have learned that an administrative meeting just ended and that food plates are still available in an open conference room. It is like a shark feeding-frenzy. Our bodies may no longer require the calories. We may have sufficient resources to purchase our own food, but the desire for free food remains strong. Our history as a species has programmed us to behave in this way: to get as much food as possible for the minimum cost [optimal foraging theory, (53).]

Our human bodies and our human metabolism remain geared for lives as migratory hunters and gatherers in the open African savannah (54; 55). As early hunters, gatherers and scavengers, we could not afford to pass by a caloric opportunity knowingly. That would be like leaving a large sum of money on the ground because we do not need it right now. Today, our modern, cyber-age bodies still love fatty foods because of this history. Humans who enjoyed eating fat got more calories on average than those who did not like fat. The fat lovers survived to be our ancestors, and they passed that desire for fatty foods onto us. We still need fat in our diet, but we do not need any more fat than we needed back then.

Our hormones beg to differ. They evolved in us during a phase when calories were a scarce commodity. Hormones steer us toward eating fatty foods by convincing our brains that those eats are desirable and tasty [e.g., ghrelin and leptin (56).] The chocolate bar, the cream puff, and the fatty burger are irresistible to most people for the same reason a rotting piece of meat is tasty to a fly, the nectar is tasty to the bee, and my flower bulbs are tasty to the neighbourhood squirrel. For each being, the food source contains a nutritional resource needed for the survival of the animal.

Our body chemistry still thinks we are on the savannah, and our brain still responds accordingly. Even when we get more than enough calories after going "hunting" at the grocery store or corner café, our brains are still on the lookout for tasty tidbits. It believes the bag of chips or the fruity candy to be vitamin-rich berries or marrowbones filled with calorie-rich fats. Our brains and our hormones never quite comprehended this fundamental shift in how we acquire our daily calories.

It is not just our brains that lie to us. Even our fat cells produce a group of deceitful chemicals. Fat-cell hormones have fancy names such as leptin, ghrelin (56), resistin (57), interleukin-6 (58) and tumour necrosis factor alpha (59). Despite their complex names, they all have a simple purpose. They regulate our appetite and how our body uses food. Leptin, for example, regulates appetite and metabolic function. Usually, when we eat fatty foods, leptin releases itself into our system from our fat cells. This hormone signals the brain that we have had enough to eat.

Resistin has the opposing role. As our body releases it into the blood, it increases our appetite. It also lowers our sensitivity to insulin. Insulin also tells our brains that we have had enough to eat. Unfortunately, resistin levels increase with body fat. Couple the higher resistin levels with the reduced insulin sensitivity in humans who have large fat reserves, and one has a vast molecular conspiracy: our bodies think they are hungry even when we have had more than ample calories (60).

The pancreas and the stomach produce another hormone, ghrelin. Ghrelin acts as a counterbalance to leptin. As our level of ghrelin rises, our appetite rises too. Just as in a crooked game of cards, this system is rigged. These hormones create cravings that trick the body into believing that it needs more calories than it does need. When we work to lose weight, the fat cells release less leptin to signal starvation. The hunger signal is a fiction. It tells a very particular region of the brain, the hypothalamus, that we will die if we do not eat immediately. Mammals, birds, reptiles and fish all possess this ancient processing centre. Among its many functions is getting us to feel hungry or sexy. The hypothalamus can spin a tale of imminent death if we do not get food RIGHT NOW. The release of the other two hormones reinforces this urgent message to the hypothalamus.

When confronted with these cerebral memos, we try to hunker down and wait until the brain and hormones stop screaming in terror. We also appeal to our sense of reason, reminding ourselves that we stand far from the brink of starvation. Sometimes ignoring the lies works. Sometimes reason works, but usually, it works only if we are already teetering on the edge of successful resistance. Unfortunately, however, the hypothalamus does not understand the concept of time and does not respond well to logic. The hypothalamus just sits back and does its nails until the thinking brain loosens its guard; then we transition from a restricted-calorie diet to a large cheese pizza in one move.

"But I Want It NOW!"

~ Veruca Salt in *Charlie and the Chocolate Factory*
by Roald Dahl

As if lies from our hormones were not bad enough, the hypothalamus learns how to manipulate us from our past behaviour. It knows when to expect food, and it learns how to manipulate the cues that surround our eating habits to control us (61). This part of the brain wants what it wants when it wants it. It has no sense of time, and it completely lacks patience. Like the spoiled little girl in Willy Wonka's Chocolate Factory (62), it wants it NOW! If it does not get what it wants, when it wants it, it throws a temper tantrum.

The hypothalamus networks with a set of brain regions known as the "reward centres" (63). In essence, the hypothalamus hijacks the reward centres of the brain and holds them for ransom. It also teams with the limbic system, the part of the brain that generates our emotional responses. If the hypothalamus does not get what it believes it needs, then it works in cahoots with the limbic system to withhold pleasure from its host until the body begins to do its bidding. It sends these messages straight to the thinking centres of the brain through a direct hotline to the frontal lobes. The message arrives loud and clear: if the body does not meet the demands of the animal brain, of the hypothalamus and limbic system, then the body will die.

Sometimes, the frontal lobes do not pay close attention to the hypothalamus. More treacherously, from the point of view of the pleading emotional brain, they may choose to ignore the demands. Then the hypothalamus coordinates a series of responses to the environment to get its way. It will throw a tantrum or push the frontal lobes into believing that there is a logical reason for the demands. The animal brain wins its case if can convince the frontal lobes to accept the urgency and justness of the case. The hypothalamus, its cronies, and its hunger signals that once kept us alive on the savannahs of the Pleistocene have now become malfunctioning lies that threaten our very lives.

When the chemistry of the brain is out of balance, the hypothalamus and its cronies become even more demanding. They can cause us to feel just plain wrong. We may feel depressed, or we may sense that we are missing something critical in our lives. Because the lie emanate from the region of the brain that controls our drives, we can develop an overpowering sense of needing something. We may experience physical or psychological pain. We may experience grief, or we may feel alone. When we find something that helps ease the pain, this part of the brain recognizes the relief and proceeds to tell us we need that relief just as surely as we need oxygen.

Anyone who has had a brush with a chemical dependency or with other addictions knows that he or she can sometimes do things to sabotage his or her recovery. It is common to hear someone in a support group say, "I don't know what happened. I went out for a newspaper, and the car drove itself to the liquor store." People find themselves with a drink, a drug, or a lottery ticket in hand, and they do not know how it got there. They are uncertain who bought the fix, but they know it was not them. They are not lying. These slips happen both when the hypothalamus and its friends function in the absence of a functional reasoning centre or when they merely override it.

The hypothalamus has some powerful partners in this con game. It physically joins with an almond-shaped region, the amygdala. This area controls our strongest emotions. The amygdala works with the hypothalamus to manipulate our feelings. Our reasoning centre tells us we feel better when we eat a tasty diet high in fruits and vegetables, but our primitive brain still tells us that we crave calorie-rich foods high in fat and sugar.

Centuries ago, these chemical stories helped humans adapt. They once kept us nutritionally sound. They let us know when we ate a plant that cured a disease, gave us a boost of energy, or provided us with an essential nutrient. They would tell us if a particular behaviour would keep us safe.

Then, we got smarter faster than our brains did. Once we learned to distil the alcohol from ripe fruits, our brain chemistry could not keep pace. Our human ancestors may have sought a slight alcoholic buzz from fruits because it meant that berries were ripe and full of nutrients. Only after we learned how to gain easy access to large amounts of alcohol could the desire for such a by-product become maladaptive.

Similarly, as we learned to concentrate the juice in the coca leaf, someone forgot to send that memo to the hypothalamus. The slight high from the coca leaf did not present the looming danger posed by our modern distilled cocaine. Our frontal lobes understand that we should not overindulge, but our frontal lobes do not always drive our behaviour.

Over the eons, animals evolved these chemical signalling systems to support their survival. The ability to transmit information within the body depends upon those systems. We have learned how to enhance or diminish these signals by artificial means. The chemicals we use for drugs mimic the chemicals in our body, and we respond to the lies they tell. This mimicry is beautiful in its simplicity. In principle, it is not different from any other form of mimicry. An array of compounds blend into the chemical environment of the brain rather than into a physical world of competing life forms. I will talk more fully about the chemical lies of drugs in Chapter 11, but for now, let's explore mimicry more fully.

Chapter 4
Lying Through Mimicry

"Bees do it"
~ "Let's Do It" by Cole Porter

Jake felt uncomfortable in his position. The tunnel under the leaves was cold and damp in the early spring drizzle. He worked at keeping the muscles in his body relaxed, knowing that it would allow him to remain there virtually motionless for a longer time. He balanced his rifle before him on a tripod, with just the tip at the edge of the leaf pile. His camouflage worked so well that even his commander could not spot his location. As a S.W.A.T. sniper, his life, the lives of his team, and the lives of the hostages depended on his ability to avoid detection.
~ "Sniper Jake" by Circa24, unpublished

As animals, we use much of our communication to convey false information. Like Jake, the sniper, many animals depend upon camouflage to convince their enemies or their prey that they do not exist.

The idea that humans are the only species capable of telling lies is a myth. Humanity has an amateur status when compared to some species' ability to deceive. For some species, their very form is a lie as they advertise an absence or a false identity to the world.

The best camouflage artists in the world include the arthropods, especially the insects. People who live in the Midwestern and

Northeastern parts of the United States and Canada often lack an appreciation for how many medium-sized and large insects surround them. They seldom see the myriad of moths, mantids, katydids, and walking sticks because these insects blend so well into their backgrounds.

Schools of swimming fish provide good examples of simple, misleading colouration-camouflage and collective mimicry (64). The silver bellies found on many fish that hang out by the surface of the water help them to blend into the surface of the water. A school of such fish can look convincingly like a bunch of ripples. Often, the fish that spend their time near kelp or beds of other vertical plants have vertical stripes to help them disappear into the surrounding vegetation. Jointly, their behaviour can also reinforce the mimicry because it can mime the movement of the surrounding area (65).

Camouflage goes beyond simple colour matching. Both patterns and behaviour help an animal to blend into its background. Lawyers know that a client stands a better chance of acquittal if she blends into the habitat of the jury. The more she looks like a "nice" middle-class homemaker rather than a stereotypical street person or prostitute, the more likely it is the jury will see her in a sympathetic light.

Animals also use this ruse. Numerous insects, such as the katydid, disguise themselves as plants. They have fine-tuned this trick for millions of years, dating back at least to the middle of the Permian era (66). The katydids—or bush crickets—are relatives of the grasshopper. They occur on every continent except Antarctica. In the Midwestern US and Canada, they are plentiful. Some temperate katydids camouflage themselves as dried-up old leaf litter in shades of grey and brown.

Others resemble a light green leaf or a leaf entering autumn colours. Their disguise allows them to sit unnoticed in a tree.

Katydids not only look like leaves, but they also behave like leaves. They move slowly through their environment. Their first line of defence against a predator is not to fly away, but to remain motionless like a leaf (67). When at rest among the leaves of their host trees, they even position themselves in such a way as to resemble the correct position of a leaf relative to the branch.

The stick insects take mimicry to the next level. They simulate sticks so well that biologists call them the *Phasmatodea* which means "phantom" insects. People seldom see these ghosts of the treetops, even in regions where the insect remains relatively abundant. When phasmids move, they gently rock as if they were a leaf or a small stick swaying in the breeze. The phasmids even rest on the underside of leaves, sheltering their disguised forms from common potential predators like birds.

I Ain't Necessarily So

~ A variation from Porgy and Bess
by George Gershwin

Despite the apparent simplicity of their deception, the rocking behaviour of the stick insect requires numerous anatomical correlates. Today's walking stick coordinates its vision and balance with its delicate movements. The movement, the camouflage, and the visual coordination evolved together as a unit. Each change acts as a pacemaker, pressuring the development of corresponding changes in the others.

It is hard to say if one of these changes preceded the others. Traditionally, biologists thought that anatomical traits drove the development of corresponding behavioural traits. However, we have learned that behaviours can also drive changes in anatomy. For example, many salt-water fish swallow water when threatened. As far as we know, it does not serve any significant purpose. Like some human behaviours, such as twisting one's hair when bored, this gulping does not seem to serve a clear, adaptive purpose. The puffer fish, *Tetradons*, on the other hand, have taken gulping to a new level. They use the behaviour to inflate their tiny bodies quickly to many times their normal size. They accomplish this feat by quickly gulping large quantities of water. Here is a clear case of a behaviour that served as the primary pacemaker for the cascade of anatomical changes that led to the modern puffer.

Many mottled, drably coloured moths camouflage themselves as tree bark. The pattern of the mottling resembles the bark of the trees upon which they rest. They also have an instinctive behaviour pattern that tells them to orient their bodies so that their markings align with the markings on the host tree.

For these little moths, it is not enough to look just like the model. They have to *behave* like it too. For the bark mimics, that means holding their wings in the right position to blend into the bark and its lichens. Because many predators look for movement, they must remain nearly motionless. Their resting position on the tree must be at just the right angle for the pattern to align with the marks on the tree. If the markings are not at the same angle as the markings on the bark, they will be easy pickings for predators, no matter how outstanding their physical mimicry.

The camouflaged insect also has another problem. She depends upon her ability to match its background. If the background changes, then the insect will become visible. The associations between insects and their hosts have evolved over millions of years. Sudden changes in their background can spell disaster for these insects.

The most famous case of environmental change and camouflage mimicry occurs in the British peppered moths. It has become the textbook example of selection pressure since its discovery by Henry Bernard Davis Kettlewell (68; 69; 70; 71) (72).

The peppered moth comes in two varieties: a dark moth with light markings and a light moth with dark markings. Birds and other predators find both equally tasty, but they can only eat the moths if they find them. Before the industrial revolution, the light moth was the norm. In the clean, pre-industrial air, the bark of the trees was light in colour, and the moths blended well into that background. Birds occasionally missed the dark moth, allowing it to survive to adulthood. This stroke of good luck was the exception rather than the rule.

As the smokestacks of industry began to appear around Britain, soot gradually darkened the bark of the trees. Now, the light-coloured moths stood out against the background. The dark moths blended into the darker background. Birds differentially preyed on the light moths because they were easier to find against the dark background. Within a short time, the ratio of dark to light moths reversed.

In the 1970s, Britain made a concerted effort to clean its air. They placed scrubbers in smokestacks to remove the soot, and their dependence on coal decreased. As the air cleared, the tree bark once again became light, and the ratio of dark to light moths again reversed.

Light moths gained camouflage again, and dark ones again became bird food.

All the World Loves a [Mime?]

~ variation on "Be a Clown" by Cole Porter

Michael Harris: "You know what they say, 'All the world loves a mime.'"
Dick Loudon: "No, it's, 'All the world loves a clown.'
People try to run over mimes."
~ *Newhart* [TV show] created by Barry Kemp and Sheldon Bull

Looking like bark is one way to dupe your prey. Looking like scat is another. Even though many birds enjoy a tasty insect or spider, bird droppings rank very low on their list of tasty treats. Some tiger moths (Arctiidae), leaf-miner beetles, and several spiders have either larvae or adults that look like bird droppings. Another insect, the spittlebug, lives inside of a ball of its own frothy waste. This goo looks like a ball of saliva stuck on a plant. Neither of these models arouses the interest in the usual array of predators that feed on the insects.

Despite the elaborate lies of the mimics,[3] the duped predators often manage to survive and to thrive. Some hunters have developed abilities that help them to see through the deceptions some or all of the time. Other predators seem to understand that mimicry exists, and will approach a new prey with caution, taking a small taste before proceeding to chow down (73).

Sometimes an animal learns how to recognize a mimic, and at other times, it requires a change in its DNA. A dupe species may need only a minor genetic tweak to recognize the mimic. However, there is the usual

[3] For a full definition of mimicry see Appendix B.

catch. For a dupe to pass on the ability to distinguish between the real model and the mimic, he or she must survive to reproduce. The duped predator must have an alternative food choice, or it will starve between meals. A duped prey must survive predation often enough to maintain a breeding population. Again, as the mimic becomes a better liar, the dupe must become a better lie detector.

The cowbirds of North America understand this conundrum—if not consciously, then at least in their behaviour. Cowbirds are nest parasites. Because they follow herds of cows or bison, they cannot nest in a single location. They depend upon other birds to rear their young. Cowbirds, like the European cuckoo, will lay their eggs in the nests of songbirds. The genetics of these songbirds program them to favour large, healthy offspring. They do not have a program to recognize the changeling snuck into their nest by the cowbirds, though some of them do have the ability to learn.

Some species of hosts can occasionally recognize the cowbird egg as a foreigner. When spotted, they toss the cowbird egg from the nest. Recognition of one's own eggs should be a significant advantage for a songbird. Camouflaging the nest or moving it beyond the reach of the cowbirds also benefits the host (74). Trickery and deceit help the nest parasite to prevail (75). In some cases, animals can gradually increase the chances of successful trickery by recognizing, or even breeding chumps (76).

Some cowbirds (77) and cuckoos (78; 79) can back their deceitful intentions with clout. They run an avian protection racket that would make the most vicious Mafia Don wince. After they lay their eggs, the cowbirds watch the nests of the host songbirds. Should they see the host bird remove a cowbird egg from the nest, the cowbirds fly into action.

The first chance they get, they swoop down on the nest and destroy the clutch laid by the host bird, and often the nest itself.

This action may seem like pointless cruelty, but from the perspective of the cowbird, they gain two significant adaptive advantages. First, the action discourages the genetic survival of those hosts who may be better at spotting and destroying cowbird eggs. An even more significant advantage for the cowbird happens almost immediately. The songbird, if it hopes to reproduce, must start a new nest and clutch. By destroying the nest, the cowbirds get one more chance to lay eggs and move genes into the next generation. Some cowbirds even use this tactic on successful nests. They smash the eggs that their hosts have laid, thus forcing them to re-nest with that cowbird's egg (80).

Mimics and their dupes evolve together. The better mimics are more likely to survive. The better lie-detecting dupes are also more likely to survive. Each species has an opposite goal. One needs to tell a great lie, and the other needs to be able to detect that lie. The result is a type of arms race. Neither species can be satisfied with the status quo. If your rival gets better at mimicry or lie-detection, then its fitness will go up, and yours will go down. Camouflage by mimicry of morphology, smell, or behaviour, however, is just one trick in this deadly arms race.

I'm a Plant; No—I'm an Ant; No—I'm a Wasp!

Camouflage is great. You can hide from all but the sharpest eyes. This shadow world also has a flipside. Another effective way of avoiding a predator is to make it think you can harm it. If an animal thinks you are dangerous because you look like a dangerous animal, it will usually leave you alone. This kind of defensive mimicry is called Batesian mimicry. A Batesian mimic will give the impression of being something dangerous or nasty, but it does not really have that property which makes it nasty.

Most birds will avoid eating monarch butterflies after their first encounter because monarchs make them sick. The predator will vomit because as larvae, the monarchs fed on milkweed. Alkaloids present in the sap of milkweeds makes most animals ill, and the monarchs—and many other milkweed insects—retain this toxin in their bodies. Rather than hiding, the monarch advertises its noxious nature to its neighbours with a bold orange and black pattern. Many insects that feed on milkweeds also share this colour scheme. This Müllerian mimicry, which is a pattern shared by harmful organisms, helps to reinforce the warning pattern in prey.

Among the organisms that do not feed on milkweed, some butterflies will mime the orange and black pattern of the monarch. They take advantage of the birds' reluctance to eat anything with monarch colouration. Other examples of Batesian mimicry include various flies and some ash borers that look like wasps but are unable to sting, and the wing pattern of some large moths that look like birds of prey.

These kinds of mimics have one big problem. Their ability to dupe depends upon the relative abundance of the model relative to the mimic. If the frequency of the potentially yummy copycats gets too high, or if the model becomes scarce, the dupes might cease avoiding the mimics. The more frequent the mimics and the less frequent the models, the more likely a predator will chance the encounter. For example, predators respond to the warning colouration of an orange and black king snake only if there are coral snakes in their environment (81). If an environment does not have these poisonous models, an orange and black mimic will gain no advantage if it mimes one.

Similarly, when a noxious species is new to an environment (i.e., has not been there for very many generations), then the other species will have had little time to evolve a mimicking appearance. California, for example, has few mimics of the poisonous pipevine swallowtail butterfly because that model has only recently immigrated to that state (82). Mimicry has not yet had time to take root.

This information lag has some severe consequences. Environmental changes have impacted the rates of migration and extinction. As surviving organisms move into new habitats, we expect the rate of the evolution of mimicry to follow it (83), but we cannot expect their rates of change to be identical. Mutations do not happen because of environmental change. Instead, the environment favours the survival of different random mutations. The initial mutation that allows the organism to create the false signal (or recognize it) may not happen right away. Once established, organisms with the trait, be they better mimes or less credulous dupes, would have a substantial advantage over animals that lack the same set of skills.

Once animals can receive the information and interpret it, nature will continue to fine-tune the signalling system (84). When a new danger or potential prey arises, the animals understand that danger or opportunity based upon biases inborn in that species. They must first perceive the signal, and then they must have a cognitive bias (brain function) about how to respond to the signal (85).

Rats and humans, for example, readily associate nausea with a food recently eaten, even if the food was not the cause of their illness. Nature programmed us to err on the side of caution, avoid poisons by avoiding foods that *might* make us sick. This same response becomes problematic for people who undergo medical treatment that promotes nausea because our brains still make the same connections. We easily learn to avoid that suspect food because it upsets our stomach. We do not learn to avoid the food if we receive a punishment in the form of a loud noise.

The flip side of this mechanism creates a positive response to stimuli that reward us. We readily form attachments to foods that give us a high boost of calories or needed nutrients, or to substances that dull pain (86). Often, we respond to common markers that alert us to the possible presence of these nutrients rather than the trace nutrients themselves. Thus, table sugar does not only dupe our brains into believing that we are eating nutritious fruits; we can actually prefer the sugar to fruit because of the concentration of this sugary marker. The chemical pull of sugar is so strong that with it, we can overcome our aversion to bitterness and enjoy the buzz afforded to us by coffee, chocolate, and tea.

To get their duplicity to work, mimes depend upon these same kinds of channelled behaviours in the suckers that they target. The method of mimicry must match the biases in their marks. The reddish fur of a tiger might make it seem more visible to the human eye, but the principal prey of a tiger is not a human. The stripes of the tiger will look more like the background to any species whose colour vision does not emphasize the reds than it does to a berry-loving human. Any organism that does not see red, or whose vision is shifted away from red (red weakness) will perceive a series of stripes that look like many things in the background. Prey species will experience chronic stress because so many things can look like a tiger, from tall plants to hanging vines. This incomplete mimicry of the tiger forces the prey to remain alert, causing sensory fatigue for the signal "striped."

The larva, of the Pipevine Swallowtail (*Battus philenor)*, use a more dynamic trick. They can change their black markings to red when temperatures get too high. The red reflects more heat and helps cool the caterpillar (87). The colour change can have three other benefits for the animal. It reduces the need to stop feeding and take refuge from the heat. It provides an honest red warning signal about its toxicity to many species. It can also reduce its infrared (heat) signature, thus making it less visible to heat-seeking predators. All insects have heat receptors on their antennae. The pipevine larvae would thus continue to "look" blackish to anyone looking through the eyes or antennae of an infra-red hunter.

For insects, the supplementation of a physical mime with a behavioural element represents a creep forward from simple physical mimicry (2). This second kind of deception combines appearance with behaviour. Here, the insect's coordination between its physiology, anatomy, and behaviour is primarily innate.

Most insects care very little about fooling humans. For most of them, we don't even make it onto their little antennal radars. They want to trick their primary predators, the insects, spiders, and other beings that prey upon them for a living. The predatory arthropods that concern these mimes have brains and visual systems that evolved independently of the vertebrates for hundreds of millions of years. Notwithstanding this gap, these impersonators often fool us too. The ability of many visual mimics to fool so many different predators, from insects to birds, suggests that despite our great evolutionary distance, insects see many of the same visual cues that we see, and they make similar interpretations and responses.

Many insects also use their behaviours to fool predators with signals that we cannot perceive. The pipevine swallowtails (genus *Battus*) feed on pipevine plants as larvae. As adults, they announce that they are inedible by producing a distinctive clicking sound. This sound is audible to their predators, although not necessarily to us. Several toxic organisms share this clicking behaviour, and predators may understand it as, "Don't eat me or you will vomit" (88). Some butterflies have also developed the ability to produce the clicking sound of the poisonous pipevine. Other swallowtails have colouration similar to that of the pipevines, but they may not make the clicks. Each of these differences tells us that the selection pressures on these populations, their predators, or their hunting strategies differs among the mimics.

Some groups of insects use a range of tricks. For example, recall our leaf-mimicking katydids. Several katydids have branched out. They have shifted their model to some large black ants. Ants are relatives of the wasps (which I will talk about later).[4] They have features that make them less than desirable for most predators. They taste acidic. They can have a painful bite. They move in large groups that will defend their territory as a unit. Many are aggressive biters. In short, ant mimicry is an excellent defence against many predators, and numerous insects reap benefits from replicating their appearance or smell (89). Ant

---

[4] Ants, bees and wasps are all members of the order Hypenoptera, or membrane-winged insects. For simplicity, I will sometimes just refer to the order as the "wasp family." There will be ones among you who complain that "hymenoptera" is an order and not a family, or that "wasp" is a not an archetypical animal within the order. To them I say, I'm not using the term in as a systematist. Instead, I want the paragraphs to be understandable.

mimics include not only harmless animals who want to use the ants as camouflage. They also include some predators, including parasitoid wasps, who mimic the ants to infiltrate their nests (90).

As social insects, ants pose an exciting target for potential mimics. They can pose a danger to predators because they act as a coordinated unit as they attack threats. However, also because of their large numbers, mimics benefit from a high ratio of models to mimics. A high density of models combined with few mimics means that predators will likely err on the side of caution and assume they are ants. Predators of ants, such as some small spiders or wasps, can use this same deception for the same reasons. When achieved, ant mimicry provides such phenomenal benefits that it has evolved at least 70 times in organisms as far-flung as spiders, beetles, and bugs (91; 92; 93; 94), with twelve known independently evolved mimics of army ants just within one subfamily of rove beetles (95).

Ant impersonators, from rove beetles to spiders, have developed the narrow waist of the ant (95). They also mime other cues such as colour patterns, including ultra-violet (96). Many of these ambush predators and nest parasites bring their stealth to the next level, behaving, and communicating like ants.

Ant-nest beetles, for example, are polyglots. They can make caste-specific sounds to communicate with the different ants within the nest, and use these sounds to fine-tune the conduct of their hosts. By making the sound of a queen, for example, an ant-nest beetle (*Paussus spp.*) or a caterpillar (*Maculinea rebeli*) can trick the members of the nest into feeding it as they were a queen, thus turning the ants into their unwitting slaves (97; 98).

Although looking, sounding, and behaving like an ant help the mimics, they also need to sound right and to smell right, or at least not smell. Some stealthy mimics evade detection by masking or reducing their odours (99). Others produce chemicals signals that convince the scent-conscious ants that the mimic is indeed a nestmate. The red wood-ant alone has twenty-two known species of arthropods that use the "smell like an ant" ruse to mislead it (99).

Unlike humans, ants have a splendid sense of smell or "chemo" sense, and information received from their eyes had better be eusonent with their sense of smell. When the scent and visual signals conflict, either smell will win the struggle for attention or the ants will attack the sender of the mixed signals. Predators thus prefer to use the olfactory channel of communication to control the behaviour of ants.

Even plants will defraud ants with a smell. An Australian Acacia, the red-stemmed wattle, infuses its seeds with a chemical signal that mimes the scent of a preferred insect prey of a local species of ant (100). The ant thus gets tricked into collecting, dispersing, and even planting the seeds of the plant.

Some species of ants often receive full credit for domesticating aphids. These ants use these little bugs like cattle, milking them to collect the sweet honeydew that they produce as a waste product. (Yes, those ants eat aphid poo.[5]) However, successful domestication often happens to the mutual benefit of both species, and sometimes, the initial move toward mutualism comes through trickery. Those aphid "cows" benefit from the attention they receive from their protective ants. The aphids produce a perfume (cuticular hydrocarbons) that make them

---

[5] Ok, they eat aphid poo. However, as a species that has a multibillion-dollar economy in bee vomit (i.e., honey), we stand on shaky ground when we judge them.

smell like their attending ants (101). The ant-aphid relationships thus may have begun with the aphids trying to hide. This chemical signature may have started to protect them against ants. Ultimately, that perfume may have helped the aphid domesticate the ant when the aphid first tricked the ant it into being its protector.

At least one species of aphid has turned on its partner. One species of aphid uses the same chemical trick to gain access to the nest of the ant. There, like a teensy-weensy vampire,[6] sucks the fluids from the ant larvae rather than feeding on a plant (102). Only some members of a clone of aphids will become vampires; the rest continue to feed on the plants and be nurtured by the ants.

Ant mimics thus include both insects and spiders that benefit from the mimicry in a variety of ways. Although many species gain protection from their predators, others are proverbial wolves in ants' clothing. When combined with other sensory impersonations, the ant-like appearance can allow a hunter to approach a group of ants without instilling a potentially deadly panic (103; 104).

Like all mimicry, ant mimicry will only be useful if the dupe sincerely believes that the impersonators are ants. Predators of ants also use a combination of signals to determine if their potential quarry makes good prey or if it is even worth the trouble. They may also search for cues that a display of flim-flam artistry can trick a specific species of ant.

Fraudulent ants are more effective if they have ant-like behaviour as well as an ant-like appearance. Successful con artists replicate the jerky movements of the ant, with its constant waving of antennae and its

---

[6] Cool note here: The name for this genus of aphid, *Paracletus,* comes from a Latin word for a helper or comforter, thus referring to the normal relationship between the ant and the aphid. However, the species name, *cimiciformis,* means "bed bug" form, a very apt name for the transformed version of this tricky little "blood" sucker.

patterns of walking (105). The caterpillars of several moths and gossamer butterflies will wave their forelegs to mimic the way an ant asks a nestmate for food (106; 107). This forced feeding (trophallaxis) by such insect versions of cuckoos and cowbirds,[7] can starve enough of the larvae to collapse an ant colony and kill the nest as a unit (108; 107).

In British Honduras, The Pine Ridge Mountains host several arthropods, including two spiders, which impersonate a single species of ant (109). Because spiders have no antennae, these liars will also wave their forelegs to copy the antennal movements of the ants. As they move along, the spiders even dip their abdomen, copying the movement that an ant makes as it marks a trail. These tiny spiders gain protection from their predators by copying a biting ant. This sham also gives these spiders a second benefit: they can use their mimicry *aggressively* to invade groups of ants and chow down on this prey (94).

Several true bugs, including some of the assassin bugs—that become very formidable predators as adults—take mimicry to the next level. Their larvae vaguely resemble ants after hatching. What is more critical to their deception is that they behave like ants—and I do mean ants in the plural. The newly hatched assassin bugs will remain clustered together, making them look like a teeming colony of ants.

If making others think you are an ant can dissuade an attack, imagine what life would be like if you could make the world think you are a stinging bee or a wasp! The sting of many large wasps is legendary. It is painful. It can be lethal to small animals. Consequently, few predators make their living by eating wasps. There are beetles, flies, and even

---

[7] Cuckoos and cowbirds lay their eggs in the nests of other species of song birds. These host birds raise the parasitic offspring, favouring them over their own smaller babies. Some species of cuckoos and cowbirds will kill the offspring of their host.

moths that have become very effective wasp or bee impersonators. As with ant mimics, bee and wasp mimics can resemble their models in shape, colour and behaviour. Several flies rank among the most successful of the wasp mimics.

Robber flies, a group of large flies that resemble wasps, use their appearance to scare away their predators as they hunt. They are very good at mimicking the colour of a wasp, but they cannot quite compete with the mimicry of another family, the thick-headed flies. These flies are so good at their physical deception that they even have the narrow wasp waist. They fool humans so thoroughly that many a student entomologist has lost grade points because they classified them as yellow-jackets or as hornets rather than as harmless flies.

Hoverflies or syrphids—named for their ability to hover in a single spot—also look like bees. One can see these handsome flies hunting for nectar in summer gardens. Like beetles and bees, these flies are vital pollinators (110), and are among the most common visitors to fields and gardens. If you live in North America or Europe, you have probably seen them in gardens, though you may have mistaken them for bees or wasps. The syrphids have the characteristic black and yellow striping commonly found on many bees and wasps in their habitat. Most have a bee-like shape and have hairs (setae) on them like a bee. Others have the narrow waist of the wasp. Unlike bees, all flies have but a single set of wings. In the hoverflies, these wings are elongated and membranous like those found on wasps and bees. This wing-pattern increases the success of the deception. Unlike many other mimicking flies, many hoverflies have mastered the art of landing on the flower upon which they feed, just like a bee or wasp.

For all of its accomplishments, even this mime has its flaws. It cannot fly like a bee. It hovers in a single spot and then makes sudden, jerky movements. Bees and wasps do not have this hovering behaviour. True bees and wasps bob up and down. One particularly vicious predator, the practised field entomologist, can spot the difference. One must, therefore, assume that at least some other predators can also see through the ruse.

Because lone bees are not nearly as scary as lone wasps, the hoverflies that mimic bees have less of an advantage than the wasp-mimics. Fortunately, hoverflies have additional behavioural tricks up their sleeves. Syrphids prefer to hang around blooms frequented by wasps and bees. The proximity of many stinging insects makes the predators less willing to chance an encounter with them. (Recall the first principle here: the success of the mimic relates inversely to the frequency of the model). Hovers also add to the confusion by making a loud buzzing sound that warns predators of potential danger. Humans can hear some species of hoverflies from several metres away. These insects undoubtedly subscribe to the view that it is better to warn away a predator than to have to escape from one.

Other hoverflies can even mimic stinging behaviour. When caught, they will press their abdomen toward the offending predator. Although they are harmless and without a stinger, the behavioural mimicry will frighten the dupe and give it a moment's pause. That moment of doubt is all the little syrphid needs to escape.

One genus of hoverfly, *Microdon*, gains an extra benefit from its models. As adults, they are excellent bee mimics. They look like stout bluish or brownish bees. Unlike other hoverflies, they do not hang around flowers like bees, but they spend most of their time near ant nests. *Microdon* larvae mimic another fearsome insect: the ant (111). And yes, we have come full circle back to the ants.

Because it is a fly, the *Microdon* larva is a maggot. Maggots do not have legs, so they cannot physically mimic the duped ant. Instead, they produce the chemical signatures of the ants. Ants, which are not too bright, focus on the chemical signal and tend to ignore the appearance of the maggot. To the ants, the *Microdon* larvae are just another ant in the colony. This chemical deception allows the baby hoverfly access to the nest where it can scavenge resources from the ant or feed on the larvae of its host.

Because *Microdon* larvae are chemical mimics, they have a limited ability to dupe others. The only ants that they can fool are the ones that produce the same chemical signature as the maggot. Usually, hosts are limited to a single species or a single genus of ant. Any other species will treat the maggot as an invader, and they will kill it, just as they would an invading adult ant.

This last bit creates a problem for the developing *Microdon*. When it becomes an adult, it loses its protective scent. Therefore, as the time for its pupation and transformation into an adult draws near, the pupa crawls toward an opening of the nest. There, it transforms into an adult early in the morning, when the cool, dark environments make the ants groggy.[8] The fly, which is more tolerant of these conditions than the ant, high-tails it out of the nest to begin its adult life as a bee mimic.

The bee flies, of the family Bombyliidae, are probably the best physical bee mimics. Bee flies go beyond the colouration and shape used

---

[8] The technical term for this groggy state is torpor. Because they are cold-blooded, insects use the heat around them to become active. In the case of the hoverfly and the ant, the hoverfly simply becomes active at a lower temperature than does the ant.

by hoverflies. These flies also get the extremely hairy bodies of bees. They have a superior physical mimicry, but their behavioural mimicry is not quite as good. They do not imitate the behaviour of bees as well as the hoverflies do. Unlike a bee, the bee fly typically does not land on the flower when feeding. There are over 5,000 different bee flies, making them the single largest family of flies. The success of this group stands as a testament to the advantages of being a liar.

As we saw with the smelly ant-mimics, sometimes a good hoax can work without any visual aids. As humans, we depend heavily on our sense of sight. That dependence makes us biased. We believe all animals will respond to visual signals the same way we do. All animals have sensory perception. Diverse species of animals respond to a sundry of signals, and they ignore others. Different organisms hunt for different kinds of cues, such as odours, sounds, or behaviours. Behavioural mimicry can work despite the lack of visual miming.

The Sub-Saharan hive beetle, for instance, is a little nest parasite that feeds on the pollen reserves of honeybees. The beetle's hard exoskeleton makes it difficult to kill, so bees protect themselves by driving the beetles into special "prison cells" inside the hive. The bees make the prison cells from plant resins, a sturdier substance than the normal wax cells (112). The jailors then post guards to prevent the escape of the prisoners. Occasionally, the beetle will extend its antenna out of the cell, making contact with its keeper. Most of the time, the guards will behave aggressively toward the inmate, but occasionally, persistent begging results in the gift of a drop of honey (113). The beetles can thus survive for a long time in their cells with the hope of escape later. By using this tactic, 1.6% of these beetle larvae survive and escape the bees to reproduce (112). That seemingly low survival rate suffices to favour the continuation of the miming behaviour because the other beetles, who did not beg, would have died from lack of food. The key to the deception appears to be mimicking the begging behaviour of hungry baby bees. The beetles have cracked the bee code without needing to look like a bee.

Keeping up with the Red Queen

In *Alice Through the Looking Glass* (114), Alice protests the pace at which she must run in the chess game devised by the Red Queen. The Red Queen, in turn, complains that Alice must be from a very slow country. "Now, here, you see, it takes all the running you can do to keep in the same place. If you want to get somewhere else, you must run at least twice as fast as that!" the Queen explained.

Biologists know this principle as The Red Queen Hypothesis. Leigh Van Valen (115) first proposed the idea to explain phenomena in which communities of organisms change together. Each species puts pressure on its neighbours as it changes its interrelationships to them. These changes happen continually, from generation to generation, as species adjust their ability to hunt, to reproduce, and to resist predation and disease.

Everyone who wins the race to reproduce uses some biological advantage. Sometimes that edge comes from a system of bluffs and challenges, or from camouflage and deception. Within the system of bluffs, many of the neighbours wear similar masks and uniforms. This shared deception poses a challenge.

If the ratio of mimics to models gets to too high, then predators will cease to understand the warning. As the chances of getting hurt on a given encounter drop, the reward of capturing edible prey can begin to outweigh the risks of catching a dangerous one. If the ratio of risk to reward becomes too low, then predators will cease behaving as dupes. This contest—this constant coevolution of predator and prey—between parasite and host is the race of the Red Queen.

Too many mimics harm a model. Monarch butterflies, milkweed bugs, and ladybird beetles only receive mimicry protection from their flashy orange and black colouration if predators know that orange and black means that they taste nasty and make you vomit. For the ruse to work, the predator must recognize the warning signal and be unable to distinguish the poisonous prey from the non-poisonous mimic, no matter if they get their warning through behaviour, scent, or other sensory information (116; 117). Predators, therefore, look for several signals of palatability, and prey often flood them with multiple false signals. They tell the hunter that either they will harm anyone who attacks them (118), or that some other danger lurks in the neighbourhood. Brown thornbills, for example, mimic the calls of the predators of their predators (119).

Twilight and night-active species gain little protection from looking like a model. They need to sound or smell like their noxious model. Noctuid moths and pyralid moths make tasty treats for bats. Some evade bats by engaging in areal acrobatics. Others jam the radar of bats (120). Still others take a less energetic approach and mimic the sound of tiger moths that taste terrible to bats (121).

Fortunately for both the predator and the mimes, nature errs on the side of caution and usually predisposes predators toward rejecting any iffy species as a possible dinner (122). Both insect (123) and vertebrate (124) predators can recognize a number of these same signals, from visual cues to scents and tastes (125). Using these signals, the predators decide whether or not to invest their time in pursuing an organism that may punish them for their efforts (126).

The advantage of warning colouration can extend to members of the same species as well as to mimicking species. Not everyone needs to bother making the poison to make predators wary. If enough members of a single species act as a noxious model, then predators will still learn to avoid all members of that species; the rest can freeload and not expend their energy producing the poison. Take, for example, the orange and black milkweed bugs (*Oncopeltus fasciatus*). Those bugs who do eat milkweed, concentrate toxins called cardenolides. When preying mantids (*Tenodera ardifolia sinensis,* or as the bugs know them, the walking apocalypse) eat one of these bugs, they vomit. After very few encounters, they cease eating this species and other milkweed bugs that look like them. However, not all of the orange and black milkweed-bugs will feed on milkweed. Some will cheat and feed on sunflowers; these bugs will not have the noxious cardenolides as a defence. Their resemblance to their noxious brethren still protects them from mantid predation (123).

In the face of such competition, in this race of the Red Queen, it is hard to see how mimicry might first evolve. One traditional hypothesis holds that warning mimics evolved from already toxic, cryptically mimetic ancestors. A mutation in colouration could have given a small part of a cryptic population a slight advantage, allowing them to move more freely. They could thus improve their chances of breeding and reproducing. The change in colouration to one of mimicry would have given them more protection as they accessed the resources that they needed for life.

This model has several problems, including how animals might transition from the cryptic to the warning colouration. The intermediates should have the worst of both worlds. They would have lost their cryptic colouration, and they would be poor mimics of the model.

Kikuchi and Pfennig (127) put forward a refinement to the hypothesis using the coral snake as a model and the scarlet king snake as the mimic. In places where coral snakes are rare, predators attacked the intermediate forms more often than they did the more cryptic and mimetic forms (128). These attacks form an "adaptive valley" that favoured the two extremes: the better mimics and the better cryptics. Their analysis showed that the king snakes with the intermediate colouration survive without this significant extra predation if many coral snakes (models) surrounded them. When more models were present, dupes were less likely to take on the risk of the hunt. This hesitancy eased the transition from one form of colouring to the other. This same effect has since been reported in other species, including butterflies (129) and flowers (130).

One must also consider the impact of gender on mimicry. Sometimes the genes of one gender benefit more from the lie than do the genes of the other gender. Males and females often look (or smell or behave) very differently from each other. Many mimetic species with significant dimorphism have cryptically mimetic females but not cryptic males. In some species, like peacocks and swallowtail butterflies, the males advertise their presence with bright colours while the females are dull and blend into the background. In others, like the white cabbage butterfly, the male must flutter around to attract the far more sedentary females.

Many researchers, including Charles Darwin, held that the seemingly exaggerated, overly embellished, and counterproductive colouration or behaviour in these males resulted from a female preference for fancy, conspicuous males. For these males, good cryptic mimicry would reduce their chance of mating. Even if a cryptic male survived longer, he would have few chances to pass his genes for male crypsis to offspring. The extravagant appearance makes him more conspicuous not just to predators, but also to mates. In contrast, being drab helps the female fade into her surroundings and to survive long enough to lay her eggs.

In dimorphic butterflies, females usually show crypsis or cryptic mimicry. Males have bright colours. They also have a trick up their wing. The above female-choice hypothesis attributes the bright colouration of these males to sexual selection by the females, or just for a preference for males they can see. However, in many species, the bright colours also warn of poison. The warning colouration of a toxic species gives the male the ability to move more safely from place to place without becoming dinner for a predator. With warning colouration rather than crypsis, the males gain a valuable benefit: they can move more easily from female to female than they could if they had the cryptic colouration that also demanded cryptic behaviours.

Alfred Wallace, a friend, a colleague, and a contemporary, of Darwin, suggested that sexual dimorphism results through changes in the protective colouration that benefits the female rather than through her choice in males (131). For example, blending into the background, or becoming less instantly recognizable as a female, can reduce sexual harassment by males (132), thus enabling the female to control her choice of a sperm donor. This hypothesis, however, does not seem to explain dimorphism fully (133). Because males and females have different lifestyles, a combination of factors may come into play.

Kunte (134) studied the molecular phylogeny (history) of *Papilio* butterflies that had Batesian mimicry in females but not in males. He found evidence that the colouration of these females had changed to become more cryptic over time. The Batesian males retained more of the ancestral pattern. His work suggests that sexual selection alone does not always drive gender differences in mimics; the female's need to hide her identity also contributes to the differences between genders.

Be it through instinct or learning, predators that can differentiate the mimic from the model will also have an advantage. In areas where there are few truly dangerous models, the predator is more likely to attack the mimic than if more truly dangerous originals were present. As they pick off the less adept mimics, the better ones survive. At the same time, those mimic species that best simulate the model will still survive better (again, if there are enough models around). The convergence in appearance continues until the gains made by various mimics cease to provide any measurable advantage. In the meantime, the race continues.

Chapter 5

Lying Parasites and the Rest of the Arms Race

"Even educated fleas do it"
~ "Let's Do It" by Cole Porter

"Everybody lies."

~ *House M.D.* [TV Show] created by David Shore

The chemistry of lying is not limited to the human body, the vertebrate body, or even the body of free-living animals. Some parasites have learned to change their chemical signatures throughout their life cycle. In doing so, they trick the bodies of their hosts into believing they are not there. Other parasites alter the signals that the host will send to its own brain. The parasite thus turns its host into an unwitting zombie who falls under the control of the parasite (135). The host will have no choice but to survive or die to service the needs of the evil overlord.

Parasites can change the personality, thought processes, and even the level of social interaction of animals (136; 137). When it suits their needs, a horsehair worm can convince a cricket to commit suicide (138); a one-celled protist can tell a mouse that it need not fear cats (139); and a tapeworm can take a typically cautious stickleback fish and turn it into a risk-taker (140). All of these parasites trick their hosts into risky behaviour because it suits their needs.

Another kind of tapeworm (*Ligula intestinalis*) infects a type of freshwater fish known as a roach (*Rutilus rutilus*). Once infected, the roaches spend more time near the banks of the river than non-infected fish (141). By altering the brain chemistry, the tapeworms also change the swimming pattern and body type of the fish. They transform their host from a graceful, sleek swimmer to a jerky swimmer with increased succulent body fat (142). They become easy, high-caloric prey for fish-eating birds. To complete its life cycle, the tapeworm needs one of these birds as its other host. Once the tapeworms complete the bird-phase of their life cycles, they easily make their return trip to the roaches. Because these birds hang out by the river, they poop into the river and re-infect the fish below.

We do not know what is happening inside the head of a fish. The fish may be coming to the shallows because of a changed nutritional need or altered environmental desires. It could be in the shallows because it does not feel well. It may be there for no good reason. We just do not know. Under the influence of the tapeworm, the fish *acts as if* it no longer understands that the shallow water is dangerous. The tapeworm has successfully duped the fish into doing its bidding. By getting that fish to swim in the shallows, it increases the chances that its avian host in its life cycle will eat the fish, thus getting it into its new home.

Humans have a similar nemesis. The humble malaria parasite, *Plasmodium*, is a single-celled organism. To them, we are merely islands for multiplication between mosquito hosts (and vice-versa). They are only single-celled organisms, but more than any other kind of organism, *Plasmodium* has influenced the biology and behaviour of humans. Humans have developed cultural, behavioural, and biological mechanisms to avoid infection (143).

In 2015, the World Health Organization estimated that there were 56.4 million deaths worldwide; more than half of these were due to ten causes. Ischaemic heart disease (15%) and stroke (11.1%) have slowly increased their impact and killed about 15 million people. These two causes have become our biggest killers since the turn of the 21st century.

Historically, however, malaria has killed and continues to kill more humans than any single vector-borne parasite in human history. Only non-infections diseases, air- and water-borne diseases, and now HIV surpass it. In low-income countries, most of which are located in the tropics, malaria still causes 4% of all deaths. Among infectious diseases (caused by an organism), only respiratory diseases (9.9%), diarrhoeal diseases (6.7%) and HIV (4.5%) surpass the carnage of malaria.[9] The 2017 report does not give the malarial deaths for children; a 2004 report estimated malaria as the cause of 7% of deaths in children under the age of five. Those who survive malaria remain infected for life (144; 145).

Plasmodium depends upon the stealth of its vector, the *Anopheles* mosquito, for its reproduction. As in other predator and prey

---

[9] In high-income countries, the only category of infectious disease to make the top-ten list is lower respiratory infections. The remaining top causes of death are lifestyle- or age-related. Diarrhoeal diseases show last on the list in lower-middle-income countries.

relationships, *Anopheles* and its human prey have run a race for survival. This genus of mosquito includes sensitive ambush predators, keenly attuned to minute changes in the environment that alert it to the presence of its human prey. They hunt within our homes, stalking at night. We tell them where we live by the human scent that we constantly emit. Once inside, they ignore the ever-present human odour that surrounds them. Instead, they find us through the carbon dioxide that we exhale, through our body odour of our sweat. When close to us, they sniff the scent of our skin (146).

The long legs of the mosquito allow her to spread her weight over a large area to reduce the chance of the host detecting her presence. The mouth of this stealth hunter cuts into the skin with minimal disturbance. The serrated proboscis of the mosquito is so good at the stealthy, painless entry that we have modelled new, less painful hypodermic designs on it (147; 148). Most of the time, her dinner never knows that she bit him. Sometimes, the only telltale sign is the itch caused by the anticoagulant and the saliva that she left behind. That fluid also may have contained the *Plasmodium* that causes malaria.

However, as the *Anopheles* mosquito has become a more surreptitiousness hunter, we have become better stealth detectors. Through natural selection, our response to the mosquito that carries malaria has become incredibly strong. Our skin becomes more sensitive to touch, and we itch more at night (149). Our skin allows more irritants into it at night than during the day (150). We thus become more sensitive to irritation at the time we are most likely to encounter *Anopheles*, the malaria bearer. At night, we are more likely to experience that itchy feeling of having a mosquito land on us than we are during the midday hours.[10] The sound of a principle African human-malarial mosquito—*Anopheles gambia*: 150 to 500 cycles per second (Hz) with a peak intensity of 380 Hz (151 p. 62)—falls into the peak sensitivity range of human hearing during our peak reproductive years (152). Even in middle age, when most people are raising their children and helping with the grandchildren, the mosquito buzz remains well within the range of their older ears that lose their upper ranges. Women, who traditionally did much of the human grandparenting also retain hearing in the upper ranges more frequently than do men, discouraging the sneaky attack by the vector.

Despite our best efforts, sometimes the mosquito wins the battle. It bites us, injects its load of parasites, and leaves us before we even know it has attacked. We thus engage in other creative battles against Plasmodium itself. People who trace their nearer ancestors to malarial belts have undergone selection pressure to favour blood factors that discourage the reproduction of the parasite (e.g., various anemias and Duffy blood-group variations). Once in such blood, the plasmodium may enter the red blood cells to set up housekeeping, but the cell itself will then attack it after it is snug in its bed.

Malaria, our arch nemesis, does have an Achilles' heel. It needs a mosquito to breed successfully. It needs a specific mosquito to get from one vertebrate host to the next. *Anopheles*, the kind of mosquito that can carry malaria, tends to be active during the night, the time we are most sensitive to the presence of the winged pest. Humans, and many other of the mosquito's favourite meals, tend to be most active during the day. The birds and mammals upon which these mosquitoes feed often

---

[10] Night is when the malarial carriers stalk us. Different mosquitoes are alert during the day.

burrow, hide in their nests, or go into their houses when *Anopheles* is most active. That behaviour is good news for the humans and other vertebrate hosts, but bad news for the *Plasmodium* and the mosquito.

Over the centuries, *Anopheles* mosquitoes and malarial *Plasmodia* have forged an alliance. Once it is inside a vertebrate host, the plasmodium begins its task of reproducing in the liver and the bloodstream. When it is ready to re-enter the mosquito, it begins to alter the vertebrate's biology. It enters the red blood cells and begins to prepare for the new home in the mosquito. When ready, the *Plasmodia* burst forth synchronously into the blood of the host. The resulting mass destruction of the red blood cells triggers chills, then a high fever, sweating, and heavy breathing. Carbon dioxide levels, a mosquito attractant, increase in the breath of the vertebrate host. Rather than being warm during the day, the body temperature of the victim rises at night. This combination of increased temperature and increased CO_2 helps the female mosquitoes find their next blood meal.

The malarial germs deceive us in another way: they alter our biological clock (153; 154; 155; 156). Malaria fools the body into being warm at the wrong time. The cyclic fever that characterizes malarial infections is the outward manifestation of the parasite development. The parasite modulates the levels of specific hormones in the blood of the host, allowing it to synchronize its population, ensuring that there will be many decedents ready to go when the mosquito comes. By warming the skin, it also increases the dispersal of skin odour, helping its transportation find a suitable feeding spot.

The plasmodium has one problem with which it must contend: its own kind. Many malarial zones harbour more than one species of *Plasmodium* (157; 158). If it does not reproduce as fast as its neighbours, the mosquito will probably feast on the blood infected by other *Plasmodium* even before the poor slowpoke has a chance to reproduce. The host may be dead by then. In the presence of the more virulent *P. falciparum,* the slower *P. vivax* responds to the competition by reproducing faster (157). In other words, just by being present as the competition, the quickly reproducing malarial germ, *P. falciparum,* forces the slower *P. vivax* to proliferate more rapidly. This race, among species and within species, creates virulent malarial organisms in regions that have year-round mosquitoes.

If the malarial critter lives in a region that has cold, mosquito-free winters, then it has a different problem. If it is too virulent, it can kill its host before the winter ends. Malarias in temperate zones (with cold winters) do better if they let their hosts live. Virulent strains die under these conditions because they cannot keep their hosts alive long enough to ensure the presence of a mosquito taxi-service to the next host.

Tapeworms, malarial germs, and even human DNA lie to and manipulate their hosts on a moment-to-moment basis. Each being communicates with the purpose of manipulating the world for its benefit. So much deceitful communication taking place around us adds credence to the philosophy of Dr. Gregory House, the curmudgeonly television character (159). Sadly, the answer seems to be that everybody does lie. Animals, protists, and even plants communicate honestly or dishonestly as it suits their needs.

Chapter 6

Wired for Honesty or Deceit?

"Cold Cape Cod clams, 'gainst their wish, do it
Even lazy jellyfish do it"
~ "Let's Do It" by Cole Porter

You can find my favourite story of a deceitful parasite among the molluscs. Several river mussels, like the pocketbook mussel, the Orange nacre mucket, Broken Rays mussel (all *Lampsilis* spp.), and the Northern Riffleshell (*Epioblasma torulosa*) have modified their gills to look very much like a small fish or a tasty grub (160; 161; 162). The movement of the gills may even mime the movements of an injured animal that would be easy prey for a large predatory fish. When a fish attempts to take advantage of what it thinks will be a leisurely meal, the mussels trigger the release of tiny larval shellfish called glochidia that quickly infest the gills of the would-be predator. With this sophisticated deception, the mussel turns the tables on the predator. Instead of getting dinner, the fish becomes the host for the newly hatched babies of these finely tuned predators (163). The fish will serve as both dinner and taxi until the juveniles are ready to release their hosts and begin their new lives as independent animals.

The story of the deceptive mussel exemplifies *aggressive* mimicry. Aggressive mimics use their camouflage to capture prey rather than to avoid being eaten. Despite this goal, the principles remain the same as in Batesian mimicry—the Batesian mimic copies a harmful model, and the aggressive mimic copies a harmless one.

Aggressive mimicry can take several forms. It requires only that an impersonator have some malevolent intent toward its dupe.[11] The anglerfish uses a lure that looks like a tasty worm to snare dinner. Aggressive mimics include tongue orchids who tempt male wasps to a copulate with its petals, and snuffbox mussels that use their flesh—their fish-like "mantles"—to lure predatory fish. The colours of the praying mantis mimic the tone of the surrounding leaves. The leopard whose spots help it blend into the trees, the tiger whose stripes help it blend into the understory, and the lion whose colouring blends into the savannah grass, also fool their prey by mimicking their background environment.

Many aggressive mimics, Müllerian mimics, and Batesian mimics have their dishonesty hardwired into their appearance. They have no choice but to use the camouflage or deception endowed upon them by the genes that determine their physique.

Most of these mimics depend upon a mixture of their appearance and a set of behaviours that support that illusion. In the Indian Ocean, the Sabre-toothed blenny fish and blue-striped fang blennies, imitate the

[11] For the purists among my readers, many of the things that I lump under "aggressive mimicry" also can be split more finely among other classes of mimicry, such as sexual mimicry and ant mimicry. Keep in mind that a mimic can be categorized multiple ways.

little cleaner wrasses trusted by many mutualist partners. Not only does this small blenny look like its model, the blue-streak cleaner wrasse, it also mimes its movements.

The dance of the blue streak advertises its good intention and signals the larger fish to come to have their skin cleaned. The mimicking blenny will station itself near wrasses, and they do the same dance. However, when trusting fish arrive for a good cleaning, the blenny will bite it before fleeing (164; 165), often adding insult to injury with the venomous bites that they use for their protection from other species (166). The duped fish may attack the nearby wrasses, thinking they have betrayed its trust.

Despite the occasional nip from a blenny, the mutualist customers continue to elicit cleaning from the wrasses. Grooming releases endorphins—pleasant, reinforcing chemical signals. The pleasure derived from grooming seems to outweigh the pain of the occasional bite. Most fish return to the cleaning stations, and eventually learn to distinguish between models and mimics. In the meantime, their scam gives the blennies both access to prey and protection from predation.

Not everyone uses aggressive deceptions against other species. Just like humans, animals can use their deceptions against their own kind. The male armyworm, for example, can trick the female of his species into mating by imitating the clicks of a bat. To avoid becoming bat fodder, she will dive for cover and stay still when she hears the clicks, putting her in the perfect position for copulation. The success rate of the deception approaches one hundred percent (167; 168).

Okay, Birds Do It Yet Again

Treachery and deceit extend beyond the heavily stereotyped behaviours of the invertebrate world. Birds and mammals also engage mimicry and often give it species-specific twists. Some birds will imitate the hissing of a local venomous snake (169). Other birds engage in behavioural mimicry when they hide their nests. Others feign a broken wing to distract a potential predator (2). Similarly, a newborn deer sports a spotted coat that camouflages it as long as it remains motionless in its nest.

Because vertebrates display some of the most complex and varied behaviours in the animal world, one would expect the behavioural resources for deception to become more complex as well. The complex brains of birds and mammals allow them to complement deceitful morphology and programmed behaviours with flexible thought, deception, and planning, (2). Despite increased flexibility in behaviour, the template of the brain still sets limits on their ability to deceive and to see through deceptions.

On the islands of Australia, friarbirds and orioles often compete with each other for the same fruits. This rivalry for resources creates tension between the two groups. The smaller orioles have plumage and behaviours that converges with that of the larger friarbirds. If they fool the big friarbird, their fellow guild-members, then the orioles have fewer unpleasant encounters with their more robust neighbours (170).

The orioles thus use the ruse that science-fiction writers have made famous: an alien species that looks like us. Because of the dominance of visual cues among humans, sci-fi authors focus on visual appearance. However, to fool a non-human species, one must consider the sensory system that it uses to understand its world.

Like the clicking katydid, birds might depend solely on sound for their ruses. Some will imitate the units of song—syllables—made by their neighbours. This mimicry can help them to both communicate and *miscommunicate* their identities (171). Several lineages of birds, including mynah birds, mockingbirds, and the hookbills (like parrots and cockatoos), can accurately copy the sounds of other birds. In the wild, male and female mynahs will imitate the calls of their neighbours of the same gender (172). This impersonation probably attracts their neighbour's mate for a short sexual encounter. For mynahs of both genders, such deception can provide them with another potential set of genes—just in case the mate at home proves underproductive.

Not all of the deceptions by mynahs centre on sex. Like other animals, they will lie to gain access to a myriad of resources. Mynahs will also give the warning calls of neighbouring primates with whom they share the forest canopy (173). Such calls can distract potential predators from nests, strengthen territorial boundaries, and scare potential competitors away from food.

The presence of behavioural subterfuge does not prove conscious deception. Before we can discuss intent, we need to know if the brain of the bird is capable of conceptualizing deceit. Prerequisites for deliberate deceit must include an understanding of how the other being will respond. In other words, the deceiver must be able to understand the perspective of his or her dupe (21; 174; 175).

An organism is said to have a "theory of mind" if it can attribute a mental state to itself or others. The benefits of understanding the perspective of another provide a compelling reason to expect other species to have a theory of mind for both their species and for the thought processes of other animals. Far from being a human feature, awareness of perspective should be widespread. Mainly for economic reasons and convenience, we have been more willing to ascribe this feature to machines than to animals.

Despite our reluctance to ascribing insight to other species, the evidence for such understanding has become overwhelming for many vertebrates. We know that chimpanzees and bonobos benefit from correctly interpreting the intent of their community mates, members of potentially hostile apes, and the expected behaviours of prey (176; 177). We can see an understanding of how others feel in the empathic understanding of the experiences of other rodents (178).

This basal awareness of others seems to arise not from our large, conscious cerebral cortex, but from those regions that we share with lizards, newts, and fish. These ancient clusters of the forebrain are the ones that alert us to danger and pleasure and control the output of actions. These ganglia organize much of our world before the neocortex ascribes conscious meaning to their input.

A "mind" is thus not an all-or-none concept. Because different parts of the vertebrate brain contribute to mindfulness, gradations in awareness exist among the vertebrates. Given the advantages gained by cognizance of self and others, at least some invertebrates could have converged on this same adaptation. Among the invertebrates, the octopus displays the greatest known range of behaviours that suggest a keen understanding of others (179 p. 272).

Octopuses show several features of thinking that converge on the vertebrate system. They exhibit long and short-term memory, engage in exploratory play, and appear to recognize others. In captivity, at least some octopi distinguish among the humans with whom they interact (180). Based upon past experiences with these individuals, the octopus responds differently, as if it understood that the humans behaved differently toward them (i.e., had a theory of intent).

Even insects have some capacity for mindfulness and subjective experience (181; 182), and at least social insects can learn by observing others. When a bee sees another bee receiving a reward for doing something that bees would generally consider to be weird, it will imitate the rewarded bee (183). If it knows that it will receive a prize, like sugar water, a bee will even push a ball—a very un-bee-like behaviour (184).

Many animals use these skills to take the next step. By knowing the thoughts of another, they can choose to deceive. By knowing the intentions of others, they can avoid becoming dupes.

Humans have a well-honed cheater-detection system (185). The willingness to punish cheaters goes beyond the immediate infraction. The perception of self-serving behaviour at the expense of the group influences how we respond to the free-rider across a wide range of interactions (186). This perception of fair-share applies not just to our friends and colleagues; we also apply this judgement to the behaviour of our employers and other socially dominant people.

We appear to share this valuable survival ability and our dislike of thieves and freeloaders with many other vertebrates. Chimpanzees reward cooperative mates and punish freeloaders (187). Both chimps and bonobos will refuse to share food with troopmates who refuse to reciprocate in some way, such as grooming them in exchange (188). Monkeys will willfully mislead dominant animals to get a share of resources hoarded by dominant animals (189). (These primates thus behave like an underpaid employee who retaliates against someone whom he or she considers to be an over-paid boss.)

Among the costliest social behaviours of humans is warfare. We severely punish members of our tribes or nations who attempt to avoid military service. We reserve special monikers for his form of free-riding, such as traitor and coward. Other primates share this aversion to those who reap societal benefits (public good) without risks. During conflicts with other troops, female vervet monkeys punish the cowards who do not defend the troop by withholding grooming. The retribution may extend beyond the withholding of social benefits. Females will attack such a free-riding male (190).

Social fish also enforce behavioural norms for the collective good (191). Bluestreak cleaner wrasse will punish any fish that endangers their food supplies by cheating. They harass and chase any wrasse who eats the mucus of their hosts rather than the parasites upon it (192).

This commonality suggests that the enforcement of the public good and the mechanisms to enforce it are key social behaviours. The commonality, however, does not tell us if these behaviours arose in some common ancestor. Enforcement of the public good may be so critical that many social species could have undergone similar selection pressures to arrive at the same solution.

Recognizing the Thief

Birds, whose ancestors began pursuing a different branch of the vertebrate tree long before the dinosaurs ever walked the Earth, show deceptive behaviours remarkably like our own.

We know that at least some birds have good memories and can understand subterfuge. Scrub jays and Clark's nutcrackers, for example, have the ability to recall events that take place at a specific time or place, a skill once believed to be unique to humans (193; 194; 195; 196). They can remember where they put food and when they put it there. If they know that another bird watched them hide a particular seed or set of seeds, their ensuing behaviour will depend upon their personal history. If they have a personal history of thievery, then they will move their catch to a new secure location (197). They seem to ascribe a nefarious motive to the other bird; they understand that the other bird will take their food if given half a chance—just as they would do.

Ravens indulge in similar behaviours (198), and through experience, they can also learn whom to trust (199). Just like with the scrub jays, there are thieves among ravens. A raven will take another raven's cache of food for personal use. Ravens show that they understand the perspective of the other birds by waiting until the owners are away before raiding the goods. If the owner remains in the area, they will search other areas before approaching the cache site (200; 201). This searching behaviour may give the illusion of a random find. If the owner is a dominant individual, the thief will usually avoid the cache, presumably to avoid conflict. Food is vital to ravens, so even a single, minor theft of food will alter the behaviour of one raven toward another (202). These birds have a scale of trustworthiness and sneakiness; getting on the naughty list with even a minor offence can make a long-lasting impression on neighbours. As with humans, once one violates trust, even in minor acts, it becomes hard to win it back (203).

Usually, ravens cooperate when they find food. Foraging as a tough but cooperative gang works very well when sufficient food surrounds them (204). The lucky finder will call loudly, bringing other ravens to the feast. This seemingly philanthropical behaviour benefits the finding bird. He or she cannot mount an adequate defence of the food alone, and refusing to share food can have serious social consequences. Once called, the responding ravens will take a share of the food from the finder, but together, they can intimidate other rivals. Sometimes, however, a raven will not give the feeding cry. If the food is defensible and if other ravens do not know about the food, then she may risk getting caught and eat it by herself (200).

Like humans and other primates, ravens have yet one more trick behind their wings. They can lie. If the food is exposed, a raven may mislead other ravens away from the food. Like social mammals, ravens have a sense of fairness. If they conclude that a more dominant raven has hogged food, it can trick the piggish bird by sending calls that announce bounty from barren locations (205).

An understanding of theft by one's species provides an adaptive advantage in a wide range of caching species. Squirrels, like scrub jays, are not very social, but they do interact with their neighbours regularly. When an eastern grey squirrel notices that another squirrel is watching him, he will move his nuts farther apart and take strategic precautions to avoid revealing their locations as a pilferage-avoidance strategy (206). Squirrels will change their "vocalizations" (their verbal communication) in the presence of possible robbers, much like how humans will downplay their wealth in the presence of strangers (207).

Squirrels have an additional strategy to make miserable the lives of would-be thieves. Squirrels will hide their real stores among decoy caches, sites that they have dug and covered, but in which they have not placed any nuts (208).

Language experiments with parrots show they too can deceive intentionally. Pepperberg (209) discovered that African Grey parrots intentionally gave wrong answers to their human handlers when the

Greys became bored or frustrated with the interactions. Although one cannot be sure about the intent of another species, one can put forward a hypothesis based upon our experiences with humans. We too will "play dumb" when we feel tired or bored with a task. Often we intend to frustrate or bore the other person so he or she will leave us alone. At other times, we may use the mild deception to hide how much we know to gain an advantage. In the absence of data to the contrary, we must take the hypothesis that makes the fewest assumptions: the bird probably shares similar motivations for deception.

Birds show great social deception skills because so many of them are social or at least semi-social. Even species of birds that do not live in colonies must understand territorial boundaries and interact with neighbours as they compete for resources. Social species develop excellent deception skills because of the complexity of social living (21). Organisms may need to avoid their predators, but they must scramble to get resources from their species.

This pressure on shared resources can increase when members of a species live nearby. One benefits from recognizing different members of a flock, and from understanding the social histories developed among individuals. On a day-to-day basis, social species compete within their group for food, territory and mates. If an animal does not happen to be the alpha, then at least that one individual can take those assets from him.

Even alphas can benefit from deception. Often a lie allows one to steal a resource with less danger than a direct assault. If one arrives late to a patch of food, a carefully placed predator warning can scare off the competition, thus allowing the animal to scrounge no matter what its rank (210).

The ability to deceive higher-ranking individuals can help an organism get more access to those precious resources. The ability to deceive lower ranking individuals will help someone keep the edge he has already obtained. The better the deception, the more power it imparts to the fraudster and the better his or her competitive advantage within the group (211). Deceivers remain in social situations, and the individuals must still function within the group or among their neighbours.

One can view the balance between deception and truth-telling as a game (36). Sometimes cooperative behaviour is its own reward. In these situations, there is no need to lie. At other times, the payoff for lying is too tempting to pass. If one animal lies and the other one accepts the lie, then the liar wins. The sucker might feel indignant, but the liar still won that particular round for a resource. If both try to deceive each other, they may both lose. The benefits within the confines of the game depend upon the value placed on the relative rewards and punishments for honesty and for being a traitor. Over time, animals tend to perform in a way that has previously produced a benefit. In many social species, the balance sheet tends to tip toward cooperative behaviour, so lying is not as frequent as one might think.

Intentional deception requires a delicate balance within the community. Different species juggle these opposing forces in different ways. A modicum of planning and a bit of intelligence gives a cheating animal an edge over the poor guy who always tells the truth. The individual must know whom she can fool and who usually sees through the scam. She will balance the costs and benefits of the subterfuge. Lie too often to the same troopmate, and he will remember you as a liar. If others see the lie, her social group will remember that she lies, leading to less successful ruses in the future and possible social ostracism by those who no longer trust her (212; 213).

And By the Way, Primates *Really Do* Do It!

Primates, like many birds, have complex social organizations that have largely shaped the evolution of their brains (214; 215). They have large brains. A large brain allows social species to interpret and perform a wide range of social behaviour, and it helps them to develop social cohesion and mutual trust. Low-ranking individuals learn the patterns of behaviour of higher ranking ones. They learn how to signal ownership of a resource to a superior, and they learn which high-ranking individuals treat them fairly and respect those rights (216).

Large, flexible, innovative brains also make primates into excellent liars. These two features seem linked. The ability and frequency of conscious deception increase with the increased volume of the neocortex, the part of the brain that regulates social behaviour (217; 174). The larger the social brain, the more likely it is that an individual can remember and predict the behaviour of others. The more someone can predict others, the more she can defend herself against deception, and the better she can practice it. For social animals, the skills of trickery, like the skills of cohesion, may contribute to the drive toward intelligent, inventive, deceitful behaviours (175).

The dynamics of a primate troop depend upon responses to kinships, recognition of individuals, respect for status, and remembering who is trustworthy and who is not. They must be able to remember who their neighbours are, and be able to understand the perspective of their neighbours (218). This understanding must be present before one has any hope of conscious manipulation of those tendencies. The complexity of troop structure means that any advantage a primate can get over a neighbour can mean the difference between surviving and reproducing, or dying a sad and frustrated monkey. The ability to deceive others intentionally (219; 220) gives deceitful monkeys a tremendous tactical advantage over the monkey who is always honest. Successful lying requires intelligence and cunning strategies (221).

Chimpanzees, (222), spider monkeys, capuchin (organ grinder) monkeys, and macaques (223; 224) can also deceive through lies of omission. They can withhold information about the presence of a resource from other individuals to gain a strategic advantage. As one might expect, lower-ranking individuals engage in more lies of omission, and more deceptive behaviour than higher ranking ones, if for no other reason that they have far more to gain by keeping their resources hidden.

Other primates seem to follow the adage of academics. When you cannot directly display your brilliance (a very sexy trait among most primates), then you must baffle them with—well, with confusing tactics. Mangabeys, an absolutely adorable African monkey, will take circuitous routes to food sources in attempts to mislead competitors (225); Macaques, like humans, fine-tune their behaviours to the audience at hand, and like practiced actors, they will mislead competitors who could take resources from them (224; 226). Not to be outdone by a mere monkey, chimpanzees (227) also indulge in baffling behaviours and lies of omission. When a subordinate primate learns the location of hidden food, he or she will avoid approaching the hidden cache if a dominant primate is nearby. Before approaching the food, the lower-ranking animal waits until the dominant one moves far away or at least out of sight. He will then will cautiously hide his intention as he goes for the food—just in case someone is watching.

Chimps, like humans, have larger brains, and they use their intelligence to reduce the effort needed to lie. Rather than expend the energy on an obvious ploy, a chimpanzee can give social clues so subtle that Agatha Christie would be envious. Like humans, chimps can plan effective strategies for webs of subterfuge that unfold over time (228). All a chimp needs do is to look. A simple eye movement can direct the attention of a rival away from a desired resource or from oneself.

To further mislead their social quarry, they may combine a glance with a hesitant move toward a decoy, then stop, as if they were trying to hide their intent (229). When in view of a dominant animal, a subtle gaze or a slight motion is more likely to fly under the lie-detection radar of the observer and thus be mistaken for an honest signal. The duped animal is thus less likely to gain access to the resource. The subtle cue is less likely to give the liar a reputation as a cheat. However, the utmost advantage, on occasion, could come from making one's social superior look like a total doofus.[12]

Capuchins, which have a more fluid social hierarchy than chimps, seem less adept at deception, but they can still do it. They also can choose when and what information to share, and they can provide purposefully misleading information (230). They often open staged boxes of food when a dominant monkey is close by and thus lose food to their superior. In species with a stronger dominance pattern, a well-

[12] Apparently, just as high-ranking humans overestimate their abilities relative to lower-ranking ones, high-society chimps also underestimate the cunning of those who rank lower on their social ladder. However, with experience, dominant animals do learn which individuals they should treat with suspicion (634; 232), and cheater-detection mechanisms usually work well (247).

informed subordinate is sometimes too intimidated to approach the food at all. He knows that the alpha will abscond with it. Primates will use this same trick with humans whom they believe intend to take their food, just as they would with members of their own species (231; 232).

Monkeys and apes will engage in other scams amongst themselves. In the absence of a threat, some primates will give false-alarm calls to send their competition running for cover. Higher-ranking animals have little need for such a con—they can call dibs on anything they want. When they acknowledge the ownership of a desirable item by a social subordinate, it can benefit them in other ways (216). Magnanimous alphas reap long-term benefits by gaining the support and loyalty of their troop. Lower ranking individuals usually have little choice in the retention of resources—unless they resort to fighting or deceit. Among the tufted capuchins, a lower-status individual will use distress alarms, such as predator alarms, to scare away his higher-ranking competitors. With this hoax, he can even get a share of a yummy treat, like a banana (233) before the rivals realize they have been bamboozled.[13]

False alarm calls may be more frequent in species that live in habitats with poor visibility. There, two factors combine to make the deception more successful. First, the environment blurs the intent (234). Other troop members can never be quite sure if the subordinate lied, or if he genuinely did believe that danger lurked. Second, the cost of not responding quickly to an alarm skyrockets in such an unclear

---

[13] Although deceit has a payoff for all ranks, animals on the low rungs of the social ladder have fewer options upon which they can rely. The relationship between rank and the use of lies, rather than clout, to gain access to resources also happens in other lines of vertebrates. It has been seen in species as diverse as pigs (635; 636) and great tits (637). Although mammals and birds diverged in the Carboniferous period, roughly 320 and 315 million years ago, both retain or have converged on this same solution to survival: lying.

environment. In an open environment, troop members would have a better view of approaching predators than in a forested one, so the liars would need different strategies for deception.

Most experiments about animal deception have used food as the reward gained through deceptive behaviour. There are, however, studies of other forms of deceit among the primates. Tactical deception has a wide range of applications among other primates, just as it does among humans (235). For example, some primates malinger. They can exaggerate the extent of sickness, injury, or inability for the benefit of an audience. A monkey will perform these acts for someone whom he or she believes will render aid, often a part of his or her family group (236).

Intentional deceit may reach its most Machiavellian form when used for the maintenance or transference of political power. Most primates have a power structure well understood by members of a troop. Vervet monkeys will give alarm calls to deceive a new intruding male. The call would distract the newcomer, giving the troop time to flee (237). This false alarm allows the male members of the troop to retain sexual rights to the females. It also puts any peripheral male stragglers, if tolerated at all, in a state of chronic stress. Such stress decreases their chances of survival.

Vervet monkeys also lie by omission within their troop. High-ranking males give alarm calls more often than lower ranking males. Calling behaviour benefits the higher ranking male because it helps save the females and young within his control. Low-ranking males spot predators just as often as high-ranking males, but they often fail to give the alarm (213). Because the offspring of the troop represent few of his genes, because he has little access to females, and because the high-ranking males are often just a hassle, the low-ranking male has little to lose with his silence.[14]

Gorillas have a very gender-based and hierarchical power structure. One older male, the silverback, rules the troop and controls access to the females. Theoretically, his control is absolute, but other males do manage to copulate occasionally, thus keeping their genes in the gene pool. Because of his greater access to the females, and because he typically copulates first with any female in heat, his genes are passed differentially to the next generation. The power of the silverback permits him to force other troop members to bend to his will and his self-interest (238). In gorillas, this control of other group members means that he gains the ultimate prize: he has nearly exclusive reproductive access to fertile females. Other troop members must be quite sneaky if they are to gain any share of the silverback's power to meet their own needs.

Occasionally, silverbacks become bullies, abusing subordinate males or females. In that case, those female gorillas will use tactical deceit and misbehaviour to weaken or overthrow a disliked silverback (239; 240; 220). The abused females conspire to become ungovernable by the dominant male ruler. The more he tries to govern, the more they flout his rule and ignore his directives. This tactic weakens his authority, and it allows lower ranking males to challenge him. The females join forces to support a lower ranking male so he can then overthrow the leader of the harem.

---

[14] Work with other species faced with similar environmental challenges, such as ground squirrels (638; 639) come to similar solutions. Like other aspects of behaviour, the form taken by the dishonesty helps organisms adapt to their environments. Similar environmental challenges can produce similar adaptations, including similar lies. These data support the idea that higher levels of kinship lead to more true alarm signals.

When power shifts in ape society, the defeated alpha-ape often must leave the troop or experience the vengeance of his former troop. His presence poses a potential threat to the new leader. Among chimps, coalitions of males will impose the new order (241) or both males and females will join in maintaining it (242; 243).

In a troop of intelligent primates, individuals pay attention to social relationships, including who appears to have power and who does not. A powerful individual will receive submission gestures from other members of the troop. When he does not receive enough voluntary grovelling gestures, he can force his subordinates to signal their subordinate status, thus reinforcing his authority. If he makes too many demands, then he could have a rebellion on his hands. A smart primate learns how to adjust the frequency of these signals to gain or maintain power. Chimps, for example, will bluff subordinate members to get more public signals of submission. When other chimps see these signs, the social rankings become reinforced among the players because the observers gain insight about dominance relationships (15). Here, perception reigns supreme.

Busted: Lie Detectors and Honest Signals

If lying confers an advantage, then there is also an advantage in being able to detect and to punish the lies. This lie-detection advantage is so vital to humans that we dedicate large, specialized regions in our brain to interpreting the honesty of others (244). We interpret the intentions of our fellow humans and make judgements about their motives. We determine how we should deal with them based on those interpretations. If thorough analysis and accurate lie detection takes too much of our energy, we may settle for less reliable methods like hunches that take less time, risk, or effort (245). These imperfect detection methods will increase the probability of detecting a falsehood, even if they occasionally lead to mistaking something that is true as something false. In other words, under suspicious circumstances, we often hold the individual guilty until proven innocent.

Knowing whether the intention of others is noble or deceitful gives us an advantage over someone gullible. If we understand the idea of a bluff, we are in a better position to defend ourselves or even to take advantage of the liar. We can prevent another person from taking our possessions, our food, or our mates. We might even be able to gain territory, food, or mates.

When primates detect deception among their troopmates, their reaction often depends upon the rank of the liar. Troop members will respond to warnings from a higher-status individual more readily than to those given by a lower-ranking one. They behave as if they know that members with lower status have more to gain by giving false signals (246). As they grow older, monkeys will heed alarm signals from lower ranking members less often than when they are younger. Across their lives, they learn not just whom to trust, but what class of individual to trust and whose warnings they should meet with skepticism.

Leaders regularly monitor the behaviour of their troop members, scanning them for signs of deception (247). If a leader catches a lower ranking individual in a lie, then the other troop members will often make sure that the perpetrator suffers consequences. Macaques (*Macaca mulatta*) tell their troop members when they have found food about 45% of the time. If the troop discovers that someone has failed to call, he or she will be harassed for a share (248) or victimized with an increased level of aggression from other troop members (249; 250; 248). When females try to hoard food through such lies of omission, they actually receive less food in the end than more generous females. Among the macaques, the penalty for being caught in a lie of omission reduces fitness rather than increasing it. The long-term consequences of cheating and deceit, including costs to kin, cost in resources, and costs in negative social interactions, may outweigh the short-term benefits of the lie (22).

Chapter 7

The Case of the Sneaky Male

"Have some Madeira, m'dear
You really have nothing to fear
I'm not trying to tempt you—that wouldn't be right
You shouldn't drink spirits at this time of night
Have some Madeira, m'dear"
~ "Have Some Madeira, M'Dear" by Flanders and Swann

Sexual Sneaks

In many species, dominant males control access to females. The males will fight for control over harems of females, or over the females in territories that they control. They prevent lower ranking males from having contact with females to maximize the number of offspring that they sire. Theoretically, the weaker, less desirable, or lower status males should not leave behind as many offspring.

This behaviour presented something of a Darwinian puzzle for biologists. The genes should become progressively skewed toward aggressive, dominant behaviour as they get differentially passed to the offspring. However, we find little evidence to support this skewing toward such Machiavellian aggression and deceit. In generation after generation of deer, some bucks remain wimpy, and a few bucks control dozens of females. If the dominant males indeed passed on their genes differentially, then we should see less variation in males over time.

In contrast, males seem to retain variation across generations. In territorial species, some males effectively control prime nesting sites while others are doomed to a marginal existence. Despite the preference of peahens for beautifully tailed peacocks, variation remains among the birds. Somehow, both the dominant and subordinate behavioural types remain in the population.

The answer to this puzzle may reside in a very sexist stereotype that virtually any woman will gladly volunteer without provocation. Men are sneaky and cannot be trusted; all they think about is sex.[15] A male who cannot get a female honestly will resort to trickery. (Before you get too mad at me, remember, I did start by saying it is a stereotype. However, across species, there does seem to be a trickle of truth running through this image, so please, bear with me for a while before you entirely dismiss this idea.)

This concept has become known as the "sneaky male" effect (251). This concept holds that subordinate males remain in the population because their genes remain. These supposedly less-fit males find a deceptive way of copulating, thus circumventing the will of more dominant males. Alternatively, they find methods of copulating with females despite pressure from usually unwilling females.

However, when it comes to reproduction, does sneakiness provide a significant windfall for the non-dominant males? A scientific theory is

---

[15] Of course, men say the same thing about women, except that they say women want money instead of sex.

only useful if it can predict. The sneaky-male theory predicts that males who are not dominant can still reproduce if they can find a way to slip past the sperm monopoly and into the sex lives of the females. This hypothesis predicts that sneaks should occur in many species and many variations. Such deviousness would be a benefit if it allowed a male to mate under conditions when he would otherwise have trouble mating or where he would usually be unable to mate. Where no mating occurs, no genes get transferred to offspring. If no genes get passed on, then no tendency toward those behaviours get passed to the next generation.

In humans, the idea that a man wants to bed every female that he encounters has become a caricature. We have a love-hate relationship with such characters. On the one hand, we recognize their prowess. The high-school male who talks about his conquests can gain transitory respect among his peers. On the other hand, we detest the lout. Women detest him for how he treats them. Men detest him because he represents the genetic competition.

Polygyny—the practice of taking multiple wives— occurs in many human societies. Indeed, 850 out of 1170 societies listed in the Ethnographic Atlas (over 72%) (252) have some form of polygyny (253). Polygyny occurs because men strive to have access to more mates.[16] Typically, having multiple wives is a luxury of the wealthy, powerful, or well-connected males. Polygyny often leaves other men with little or no access to women (254). These remaining men often fight each other as disputes over women arise (255). Laws against

---

[16] I do not mean to imply a conscious, goal-oriented desire. In humans, as in other animals, the desire is largely innate and unrecognized by the person. It's more akin to wanting to lose weight, but finding yourself going through the fast-food drive-through without realizing you are doing it until you are there.

polygamy help to protect less-dominant males from losing access to potential wives (254).

The human desire for wealth and status has its roots in this kind of competition. In many societies, human males must pay a bride-price to the family of the potential wife. In polygynous societies, such bride-prices are standard (253). Males with higher status and greater wealth can afford larger households. The ability to support multiple wives reflects their high status. Dowries may be a similar form of female competition (256). Dowries given to grooms may also offset the energetic costs of starting a new household in the husband's village.[17] The woman's parents benefit from establishing conditions for healthy grandchildren more quickly.

In many social species, males will use wealth and status to attract mates. Male dragonflies and damselflies compete to have exclusive access to the best spots for laying eggs (257). Many species of male birds also do whatever it takes to control hunting grounds around a central nesting site. In these species, females look for a necessary resource and then select the mate based upon who can give her access to that resource. Even in species like aphids, which can reproduce without a male, females will compete for access to the best possible resources for their young (258; 259).

In other species, "inferior" males must overcome a different reproductive barrier. Females may want males with specific physical traits. A doe may want a buck with a big rack of sexy horns. The female warbler wants the bird with the best song. A female house finch wants the reddest feathers in her mate, but the zebra finch wants her mate to

---

[17] Women who entered convents in the Medieval era were also expected to provide a dowry. Some convents only accepted postulants who had sufficient funds.

have white downy under-feathers. The female cichlid fish wants a big male who can defend her and her eggs against all others.

In almost all species, females put much more effort into choosing their mates than do males. They look for that rack of horns, listen for a complex song, assess the territory, or watch the mating dance closely. They do so for a compelling reason. Females use these specific signals to determine the health and reproductive fitness of the male. The buck cannot fake a large rack of horns. The frog and songbird cannot sing an extended and continuous song unless they are healthy. An unfit dragonfly cannot defend that patch of ideal reeds. Females respond to these species-specific honest, un-fakeable signals as if programmed to do so by their DNA (260).

Males, on the other hand, tend to be far less fussy. They prefer females with signs of fertility, but in many species, males will court anything that looks vaguely feminine. In guppies, even a slightly rounded shape on a stick (the shape of a fertile female) will trigger a mating response. Female guppies, on the other hand, prefer males with orange spots that look like food (261). When one looks at the time-budgets and energy-budgets of these fish, the male uses much of his time and energy trying to copulate with females, while a female will spend much of her time trying to avoid unwanted copulations (262). Males spend much time moving in search of new potential mating partners (263). Females develop Olympian swimming styles, using less oxygen to swim and to outpace males when necessary (264); when combined with their ability to dive quickly to the bottom, they can effectively avoid many of the unwelcome amorous advances of the males. Why do we see such a difference between male and female behaviour?

In the mid 20th century, a biologist named Bateman (265) noted that these fussy females usually have far more at stake in mating than do males. Females produce fewer eggs during their lives than males produce sperm. Eggs take more energy to make than sperm. After a male mates, he usually can court again and fertilize more eggs.[18] The female must lay the eggs or carry the young. She may or may not have the energy for a second round of ova. One can think of females as a limiting resource in sexual reproduction (266). No matter how much sperm a male can make, there are only so many females and only so many eggs to fertilize. On average, females put more effort into mate selection, and they spend more time with their offspring. Males need to compete for eggs. Females need to choose carefully what sperm reaches their eggs. It is the female, not the male, who is the buyer in this market, and she benefits if she accepts the best offer. For many mating species, females benefit if they sneak away for a fling with a higher quality male; males, however, benefit if they pursue as many females as possible.

Sometimes, both females and males benefit from the sexual affairs of females. Males who have had an affair with a female may be less willing to attack her and her consort. After all, they may be raising his egg. The confusion can also increase the probability that a neighbouring male will render aid when the mated pair cries for communal help to fend off a predator (267; 268).

---

[18] I am excluding "broadcasters" from this discussion. Broadcasters release the sperm and eggs into the environment. Included in this category are some corals, and a few worms whose bodies literally explode to release their eggs and sperm. Be it a male or a female, an exploding worm usually has only the one opportunity to mate.

They Called Him "The Sneak"— Fastest Frog on Four Feet!

~ with apologies to Ray Stevens

Many male animals do not have a prayer of mating if they represent themselves honestly. Either the underdog must outfox the better-adapted male, or the loser must convince the female that he is a better catch than he truly is (251). When males engage in this deceptive form of mating behaviour, we call them "sneaks." Biologists define sneakers as males that appear to be poorly adapted for reproduction but who are still able to copulate with females by outfoxing other males or by misrepresenting themselves to females.

In humans, males often misrepresent themselves to females. They may claim to be more wealthy than they are, more powerful than they are, or free of communicable diseases when they are not. Mailroom clerks may pretend to be vice-presidents while at a pick-up bar. At that same bar, few vice-presidents are likely to portray themselves as mailroom clerks. Biologists have found such sneaky behaviours across the animal kingdom. Like other forms of miscommunication, sneakiness is far more widespread and far more diverse than initially believed.

Such shifty sexual tactics occur in all the major lineages of the vertebrates. Sneaking can lead to significant reproductive success. We see the behaviour in fish (269) and amphibians (270). There are devious birds (271; 272) and Machiavellian mammals (273). We even see it in our primate cousins (226). The methods used by the males to fool the females or to thwart their competitors vary with the environmental and with the perceptual demands on the males.

Who is the Real Fool?

"I doubt that female mimicry exists anywhere
outside the imagination of biologists."
~ *Evolution's Rainbow: Diversity, Gender, and Sexuality in Nature and People* by Joan Roughgarden

A generation of behaviourists has written tomes on the sneaky mimics described above. The story of a substandard male cheating a stronger or dominant male of his hard-wrought mating rights makes perfect sense. The idea that a low-quality male must surreptitiously insert his sperm against the wishes of a female neatly fits our model of choice and reproductive privileges. One small problem persists with the model. No one has demonstrated that either the "cheated" males or female failed to recognize the female mimic as a male.

Roughgarden (274 pp. 91-101) proposed an alternative hypothesis: the two recognized principles in the mating game could recognize the "deceitful" male as a male. Both the "dominant" male and the female may accept his mating.[19] In some species, a less aggressive female mimic could encourage females to enter the territory of a potentially dangerous dominant male. A less aggressive male may be related to the dominant male, and thus gain some reproductive fitness from his brother's mating in the form of nephews and nieces.

Alternatively, for the price of a few fertilizations, a dominant male may be able to improve his reproductive success because a subdominant male could help attract females, act as matchmaker, build nests, or defend resources. The genial-gene hypothesis switches our

---

[19] Roughgarden rejects the term "dominant" and "submissive" for many reasons, including a lack of evidence that social dominance is part of these dynamics in all species. The term also produces a bias in our view of these social relationships. I use the term here because we have no agreed-upon, widely understood substitute for this term.

frame of reference from a simple linear relationship between a winner and loser in direct competition for eggs, to one in which two players (the two male forms) can each increase their fitness (gain more eggs) through cooperation as compared to their efforts as separate individual (275). In many cases, cooperation, not competition, gives the better genetic pay-off (276).

The model of Roughgarden changes the frame of reference from competition for exclusive control of eggs to cooperative ventures that can obtain more eggs for the dominant male and ensure some eggs for the less-dominant male. As in a corporation, the mimics may advertise their willingness to be an employee in exchange for a fair share of the bounty, albeit a lesser share (277). In cases where the Roughgarden hypothesis works, the female may prefer the variations in sperm available through the successful partners rather than the single source of sperm from one specific male. In this model, one would view the female "disguise" as a camouflage done with a bit of a nod and a wink to both of the cooperating parties.

Sexual dimorphism among the males may thus benefit both the males and the female. If Roughgarden proves correct, then these males whom we interpret as "feminized" would not be deceitful sneakers but a different expression of gender. In that case, the only ones fooled would be the humans watching them.

Bribery

"A man's gonna sweet-talk and give you the glad eye, but when that sweet-talkin's done, a man is a two-faced, a worrisome thing who'll leave you to sing the blues in the night."
~ "Blues in the Night" by Harold Arlen and Johnny Mercer

In many species, females expect males to offer a courtship gift. Humans are no exception to this rule. Human females judge the ability of a male to provide good genes and resources for her young just as surely as do many insects and birds. Human females (and their families) look for males who can make a high investment in her young, and who are willing to make that investment (278; 279). For humans, the show of resources may be the quality of dates, a diamond ring, a herd of cattle, or a song played on a handmade flute. In other species, the gifts may differ, but they represent something essential to the female.

In many species, males offer a gift of calories. This added energy helps the female prepare herself for the task of laying eggs or raising the young. The males must invest energy into procuring these food items, but they increase the chance of survival for the females who will carry the genes from their sperm (280).

The nuptial present need not be food. It could be a different resource such as a territory for laying her eggs or a quality nest. Females of other species want to increase the cost of reproduction for the male. In "monogamous" species, this pattern is common. The higher the cost of sexual access, the less likely it is that he will leave her after mating. In promiscuous species, the demands of a female could still reduce the chance that a quality mate will have opportunities to fertilize other females. She could thus monopolize more quality sperm for her eggs.

Among scorpionflies, the females only need sperm to fertilize their eggs, not help to rear offspring. The male must present his lady with a tasty moth or another dead insect (281; 282 pp. 54, 78). This gift of food is the scorpionfly's equivalent of an expensive diamond ring. The females demand this gift before mating.

For the males, the need to obtain the moth is thus critical. They may rob another scorpionfly of his catch. Such theft takes less effort than honest hunting. Theft also forces the victimized male to hunt again before he can mate. If they get desperate or if times are rough, these male flies will even steal food from a spider to court their lady loves (282 p. 271).

For large males, obtaining an impressive moth is usually a cinch even without resorting to thievery. Even an average Joe can catch at least an acceptable fly or moth. Smaller, non-muscle-bound hunks of male scorpionflies may use a smaller insect or try to fool the ladies with a wad of saliva. Consider this spit to be the cubic zirconium of the stonefly's world. Smaller males may not even get a good wad of spit to offer her. Females usually reject this smaller calorie-free nuptial gift. Sometimes that rejection leads to forced copulation (282 p. 272).

If It Looks Like Rape . . .

Few traits are limited to humans, and rape is no exception. We also find rape in other species as well. Few things offend our human sensibilities more than rape. Females consider it the ultimate violation of their person and a form of sexual slavery. Many women never get past the experience of lost autonomy; the trauma and post-traumatic incapacitation ruin their lives. Men hate it, not just because it violates their sisters and mothers, but because it also violates their sexual partners, their wives and lovers.[20]

Not all species share our aversion to coerced sex. The costs and benefits of coercion can vary across species (283). For some species, rape provides few reproductive benefits relative to its risks. For others, forced intercourse is the reproductive strategy of last resort used by the desperate male. Other species, especially for those in which the male has a high investment in each copulation, the unfettered access to females becomes paramount to success. For these species, the removal of female autonomy carries a substantial payoff for males.

A male hamadryad, a type of baboon, will harass the females under his control throughout their lives. Female elephant seals will endure violent encounters with males many times their size despite their attempts to resist sex (284). Male guppies will continue to harass females until they find a chance to copulate (262). Females will often dive to the bottom of the river (or their fish tank) in a vain effort to avoid these painful advances. Male wagtails, a type of river-shore bird, will threaten or even attack females to force them into copulation and mating (285).

The species above do not hold a monopoly on forced copulation or on female attempts to prevent it. In damselflies, females must position their abdomens to allow males to mate. Usually, they will only mate with a male that offers proper courtship and who has the appropriate territory: a luxurious stretch of shoreline with the right mix of plants

---

[20] From the perspective of the DNA of the male, it also gives the genes of the rapist an unfair advantage. Several social species appear to have programmed their hosts (the animal), to lash out against behavioural cheats, such as rapists and philanderers.

and shallow water. The male damselflies then guard their females, making it difficult for other males to gain access to them.

In places where there are very many males, these dual barriers do not deter the hapless males who have no such territory. These males will repeatedly try to force copulations upon the females, succeeding about 55% of the time (286). Afterwards, they will prevent the female from mating again by sitting on her wings.

If there is any measure of justice for these damselfly females, we find it in the stupidity of the males. After a forced mating, the defilers frequently guard the wrong female. Their behaviour tells us that they do not recognize the female with whom they have just mated. It also reduces the probability that the sperm of the debauching male will get passed to the next generation.

Other orders of insects may also engage in rape. For scorpionflies, the attempted debauchments are not generally successful. Average-size males usually have more success if they find another female and try again. However, for a small male, the success rate for consensual sex is low. He will therefore repeatedly try to force copulation. For these small males, the only way they can hope to reproduce is through rape.

When a male camel cricket cannot induce sexual cooperation from a female, he will switch tactics. His hind legs bear weapons designed to fight other males; however, when shunned by a desired female, he can also use these weapons to force females into submission (287). The better his armament, the less attractive he seems to be to females, but the more power he has to force himself upon them (288).

From the perspective of the female, the dual threats of sneakiness and coerced copulation present a fundamental problem. Both tactics circumvent the mating choice of the female. False communication and forced copulation hinder her ability to choose who will be the sperm donor. If she can find a way to restore that choice, in full or in part, she has a genetic advantage over the female who cannot exercise choice.

Wherever it occurs, female animals dislike forced sex, just like humans, be she a camel cricket or a cuttlefish (289). Females develop different tactics to limit the unwelcome advances of males, but males counter these defences by becoming better sneaks or rapists.

Among the Japanese flying squid, a female usually lays one mass of fertilized eggs and then dies. However, she may expel a smaller clutch of eggs prematurely if an amorous male interrupts her egg-laying process; she will then extrude the remaining eggs later. Males, however, will turn this tactic to their advantage. If a female cannot repel the sexual advances of these pesky males, they will interrupt her egg-laying to force multiple spawning events (290).

Among these squid, each interruption allows the males additional access to females. The smaller egg masses have a reduced chance of survival, but those offspring have little chance of being the spawn of the offending male. He thus plays a game of reproductive roulette. In this game, the odds favour a reduced reproductive chance for his rivals and only a slight risk of harming his own young. However, because the victimized female is more likely to survive the aborted attempt at egg-laying, he has gained a potential mating opportunity. During a future encounter, he can still pirate some of the remaining eggs.

The females try to block forced sex and spawning interference by males. Males parry and find ways to exert their reproductive needs above those of the female. The Red Queen, that strategic struggle between predators and prey, becomes a player in the mating tactics of both genders across a wide swath of species from insects to vertebrates.

Waterfowl males frequently rape waterfowl females. Just as in humans, mates can force copulation (as in spousal- or date-rape), but so can another male who usually would not have sexual access to that female (as in stranger-rape). Such ravishment often happens among waterfowl. Here, non-consensual sex is so evident that we have recognized the behaviour at least since the early 20th century (291). We have seen it in many species, including mallards, African black ducks, snow geese, and pintails (292; 293). In some species of dabbling ducks and geese, females have developed an anatomical mechanism for fighting forced copulation.

To study the problem, one must differentiate between forced and consensual copulation in birds. Fortunately, we can easily make this differentiation on one key behavioural difference in ducks. Male and female genitals do not fit together easily. Females have a vagina that twists and turns to resist the penis, not to embrace it. This anatomical mismatch makes it very difficult for a male waterfowl to penetrate a female against her will (294). Unfortunately for ducks, even willing partners have difficulty with sex.

Consensual sex may appear aggressive to humans, but it habitually takes place in the water. In ducks, most acts of non-consensual sex appear to take place on land. This dramatic difference, among others, makes it easy for even the most casual observer to distinguish rape from consent in the world of ducks.

Another critical behavioural difference between consensual sex and forced sex in ducks is whether they engage in post-coital bragging. Usually, males "boast" of their sexual prowess after voluntary copulation. In mallards, males do not follow forced copulation with their typical post-copulatory display of bragging (295; 296).

Rape does not trigger the instinct to do his strutting display. We do not know why. It is unlikely that the bird feels guilty. Perhaps he wants to hide from the victim. He may wish to avoid the wrath of her mate or other males. He also may be programmed to hide the encounter because it could trigger retaliation against the already victimized female, thus impacting the chances of survival for any egg that may result from the rape.

Female waterfowl resist rape, and they often become exhausted in the process of trying to elude the aggressive male. A female may try to attract the attention of her mate, asking for aid, but there are downsides to being a visible maiden in distress. Her mate, who often will defend her by attacking the rapist, also exhausts his energy reserves.

Other males may also hear her cry for help. They can be attracted to her in her weakened condition. Then, she can become the victim of gang rape (293 p. 255). If her consort has defended her from the initial attack, his weakened state can prevent him from mounting a further defence.

In some species of waterfowl, females suffer high rates of forced copulation. Snow geese and Ross' geese have some of the highest extra-pair copulation rates found among geese. About 46-56% of all copulations are outside of paired couples. Males force the vast majority of these sex acts upon the females. Despite the high rates of rape, these copulations seldom result in offspring. Extra-pair paternity in geese is among the lowest recorded, with both Ross' and snow geese having only 2–5% of their young from rape (297). We can see a similar pattern in captive mallard ducks, where about 12% of forced copulation results in offspring, even with fertile females who can store sperm for up to 17 days.

We know of several mechanisms which could help to explain the low rate of conception during the rapes of the ducks. Many of forced copulations happen near the end of the breeding season.[21] In that time frame for migratory birds, such as the snow goose, a higher proportion of lower quality mates arrive at a breeding site, after most mating has already occurred. Thus the late breeding season can bring females into more sexual conflicts with low-quality males who must then depend upon coercion to breed.

Because of the timing of the late-season rape, these sexual acts would do little good for the offending male. Offspring do not receive genes that support attacks on infertile females because no offspring occur during these rapes. For rape to be a successful sneak-strategy, it has to occur during times of fertility. Males need to mate with fertile females, not with infertile ones. For some species, males can use telltale signs of fertility to their advantage.

The data show that the attraction to fertility is true for chosen partners and rapists alike. In captive white-cheeked pintails, forced copulations coincide with female fertility, with a secondary peak when she is in laying condition (73.5% of the rapes). In wild scaups, about 60% of observed rapes occurred during laying or pre-laying times. The snow geese females who have the fewest fertilizations resulting from these coercions were raped late, at the end of their fertile period or

---

[21] Late-season increases in coercion can reflect avian rape as a last-chance strategy. Waterfowl do not hold a monopoly on late-season cheats. Female house wrens who mate late in the season are more likely to cuckold their mates (644), possibly as a response to a perceived lower mate quality. In the migratory yellowthroat warbler, lower quality mates can also arrive later, when much of the successful breeding has already occurred (645).

beyond.[22] Because snow geese live in a nesting colony, it may just be difficult for males to get to the females while fertile (293).

Sperm has less success fertilizing the female's egg if the intercourse occurs more than an hour after egg laying. During this interval, the next egg in the clutch is most likely to be fertilized. Through this critical time, mate-guarding peaks, and the male chases away intruders with greater vigour than at other times. Male waterfowl need to protect their reproductive investment by preventing cuckolding by the female or her rape by other males. However, he also needs to help care for his other reproductive investment, the eggs and hatchlings.

The high level of behavioural monogamy or bigamy—one or two females to one male—among waterfowl makes this phenomenon even more interesting. In the species studied, the males tend to spend much of their time and energy helping to care for the offspring of their one or two mates. As early as 1910, Heinroth suggested that this system allows females an extra shot at reproducing with another set of genes (291). The partial emancipation of the female from brood care permits sexual activity to take place without great physical damage to the female from an aggressive partner or by an angry mate. It would also provide a second mate who can protect her from aggressors.

Female emancipation may take a more intense turn. In some species of swans, males benefit from forming same-gender pairs to raise eggs. After mating, they chase away a female after she deposits her egg in their nest. Two large males can better defend more eggs with greater genetic diversity than a mixed-gender couple. Because of the timing of their copulation, the male pair can have some faith that the eggs carry

---

[22] We know about the success rate from several lines of evidence, especially DNA analysis.

their genes. The female can still produce sufficient eggs for her own nest (298 pp. 168-213). By consenting to mate with one or both members of the male pair, she provides them with an egg, she avoids a possible forced encounter by two males, and she increases her reproductive chances by placing an egg in another nest.

Among the ducks, females also have another anatomical trick that helps them reduce the incidence of successful rape. Many of them have very complex vaginas. There are dead ends, blind allies, and clockwise coils. The male penis has a counterclockwise coil. The reverse coiling and false ends make it difficult for a male to penetrate deep into the female without her cooperation. In one study of 16 different waterfowls, Brennan (299) found a positive correlation between vaginal elaboration with the length of the phallus of the male. This complex anatomy increases the control of the female over insemination and decreases the probability of impregnation by coerced sex.

Because of its relative ease, we have extensively studied this race between male and female genitals in waterfowl. However, we see similar attempts of males and females to win reproductive encounters in groups as widely diverse as whales and insects. Female whales, dolphins, and porpoises have evolved vaginas designed to prevent unwelcome access by males; males of these species have evolved penises to work around the imposed limitations (300; 301). Beetles have engaged in the same battle of the genders, with the spiralled shape and stiffness of female genitals being met with countermeasures in penis morphology by the males (302).

Forced copulation in other animals shows parallels to rape risks experienced by human females. For years, the prevailing wisdom has been that all women are at risk. That anywhere between a quarter and a half of us would become victims of rape in our lifetime. Conventional wisdom saw older women as more likely to be raped than younger ones; that rape is a crime of violence and not a crime of sex.

Let me address that last point first. It hinges on a logical fallacy known as "the limited choice." A limited choice argument tries to convince the listener that one cause excludes the possibility of another cause, or that an either-or situation exists. The idea that rape must be either a crime of violence *or* a sexual act is bunk. Rape is both a crime of violence *and* a crime of sex. It reduces a fellow human being to an object, placing him or her in the violent and demeaning position of sexual slavery for the duration of the rape. It also involves a sex act. The two choices do not mutually exclude each other. One can perpetrate an act of violence on another person with a fist or with a genital organ.

Age also plays differently upon the sexual stage than we once thought. At one time, data suggested that men raped older women more often than they did younger ones. This observation would seem to fit the pattern of late-season rape, or last-chance rape, that we saw among some of the waterfowl. However, those males entered a population of females who had become infertile. They did not choose an infertile female on purpose. Older women may be opportunistic targets because of dwindling physical strength and their diminished ability to repel the attack.

However, we fixed this idea, that older women are preferred by rapists, in our consciousness before we recognized other forms of non-consensual sex as rape. Back in the 1960s and 1970s when that idea took hold, most people did not consider rape to include date rape, acquaintance rape, or spousal rape. Once we added those figures to the fray, the risk of rape among younger (fertile) women skyrocketed. The pattern of non-consensual sex among humans began to look more like the pattern we see in other species.

Human risk factors for rape include past victimization, young age, and a diagnosis of post-traumatic stress disorder (303). These factors can influence a potential rapist because they suggest a safe target, one for which retaliation is unlikely.

The societal perception of a sexual encounter, including rape, also influences its probability (304). The level of condemnation of rape by women and the surrounding community, and the inhibition of coercion through bystander intervention, also alter the level of rapes and their success (305; 306).

Minority ethnic status, depression, and drug use also impact the rate at which rape happens. These factors link strongly with an increased risk of *any* vicious assault, including rape (303). Although these measures correlate with a higher risk of rape, other risk factors correlated more strongly with rape than with other results. The only high-risk factor that was common to sexual and nonsexual assaults was past victimization.

That does not mean that drugs carry no importance. They are just not *the most* critical factor in this case. The use of mood-altering drugs, such as cocaine or alcohol, does increase the probability of becoming a victim of sexual or other physical violence (307). One American study found that 62% of the women who used crack cocaine reported suffering a physical attack, and 32% reported being raped (308).

Conventional wisdom holds that a woman who drinks to excess at a fraternity kegger has an increased risk for sexual assault. She does. Drugs, including alcohol, do factor into the probability of rape, but our taboo about looking at the drug history of the victim (because it feels like blaming the victim) makes this type of research more difficult.

In 2003, approximately 9% of high school students reported having forced sexual intercourse (309). At least one in six women and one in thirty-three men in the United States have been victims of rape or attempted rape in their lifetimes (310). In other words, you can be a sexual scumbag or a victim of sexual assault no matter what your race or socioeconomic status.[23]

You may wonder why I included a chapter on rape in a book about deceit. Ethologists—scientists who study animal behaviour—label forced sex a "sneaky" sexual tactic, be it in scorpionflies, geese, or humans. Hence, rape is part of the greater lie. Also, lies or seduction often precede the attack. Males must get close to a prospective victim. Because females may actively avoid these males, they use sneaky tactics in their approach. Also, the male kin and consorts of the females may attack them, so they use guile to avoid detection. In that sense, forced copulation resembles other forms of sneaky behaviour.

Just because a behaviour is "natural" does not make it good. Arsenic is a natural component of mineral water, but that does not make it something you want to drink. Under no circumstances does "it happens in other species," justify a harmful or hateful behaviour. Humans form our codes of behaviour to reinforce our awareness of what actions will be tolerated and what will be shunned. Because of their long-lasting impact on the victims, their consorts, their families, and society, we

---

[23] Gathering data on human rapes is difficult. Studies often use police statistics. These statistics only can include the reported rapes. Many people fail to report spousal rapes and date rapes. Not every victim labels his or her sexual assault as a rape. Many women hide their rapes for many different reasons. Male victims of rape, we believe, seldom report their assaults.

treat this particular form of sneak with a contempt that few other behaviours merit.

Fifteen Frogs on a Dead-Frog's Clutch

with apologies to Robert Louis Stevenson

Males will sometimes co-opt the eggs of the female in a far less direct manner. They do not take sexual advantage of her. They take advantage of her eggs.

Males of many nest-building fish, for example, will steal (or agree to raise) fertilized eggs from another male. They deposit these eggs in their nests to attract females (311; 312; 313). The females interpret the presence of eggs as a sign that the thieving male is a good catch.

Sometimes in nature, a behaviour that evolved for one purpose develops a second, different purpose. Anoles will nod their heads and do push-ups as a warning to predators and as a dance to attract mates. Rhinoceros and stag beetles use their large horns to fight each other and to guard their territories. Fish gulp water to increase the flow of water over their gills. Pufferfish have modified that behaviour to inflate their bodies. Modifying behaviour from one purpose to another is so common that it should not surprise us when such modifications appear in a sexual context.

When frogs mate, the male typically climbs on the back of the female and guards her against approaches by other males. The frog will clutch the female with his arms in very tight grip called an amplexus. He will use his rear legs to kick other males and to keep ownership of his prize: the female and her eggs. Without those eggs, his sperm will not make it into the next generation. His DNA would be at the end of a continuum of successful matings that has endured for nearly a billion years.

Biologists call an innate behavioural pattern, like that of the male frog, a fixed action pattern (FAP). Once triggered, a FAP tends to continue until the animal completes the last action in the sequence. In this case, the male frog stays attached to the female, grasping her until eggs have emerged and he has released his sperm to fertilize them. He gets that chance to fertilize her eggs only if he can maintain his grip. His genes then get to the next generation, and all is well with the world.

Most frogs lay their eggs in a "clutch" or a cluster of eggs. When the female releases the eggs, the male responds by releasing his sperm. One would think after his ejaculation the result would be a done deal, but the story is far from finished.

Until 2004, this story seemed clear and precise. Then David R. Vieites and his team (314) reported on the behaviour of the European common frog, *Rana temporaria,* in a population found high in the Pyrenees Mountains that separate France and Spain. There he found many males singing their songs and competing for the attentions of the far too few female frogs. The skewed sex ratio was four to ten males for each female. Few males got a chance to mate and to engage in the fixed action pattern of mating.

We hold a common misperception that the first sperm to reach an egg gets to fertilize it. One comic from *The Far Side* series by Gary Larson (315) even showed a sperm with an outboard motor laughing and taunting the others as he outpaced them.

In many animals, the first sperm to reach the eggs destroy themselves as they try to break through barriers of slime and protective cells. Only after the initial onslaught do the lucky ones arrive and make their way into the eggs. This scenario holds for frogs too. When the male fertilizes a clutch, his sperm will attempt to penetrate the layer surrounding each egg. The first sperm on the eggs will not be the first sperm cells that succeed. Instead, the sperm behave like an invading army. The ejaculum sacrifices pawns so others of his sperm can reach the egg.

These sacrificial sperms pave the way for later sperm to enter, even if they are not the sperm of the original donor. Male lampreys, for example, may wait near a mating couple. Among these fish, fertilization is external, after the eggs leave the body of the female. As the mating male squeezes the eggs from the body of his lady love, a lurker may swim toward the couple and ejaculate over the eggs (316).

In the Pyrenees Mountains, male frogs can use their innate fixed action mating behaviours in a novel way that takes advantage of this same pattern. In this sneaky form of mating, eggs alone, even without the presence of the female, triggers the fixed action pattern for mating. This second male typically claims the clutch within seconds after the female lays it, grabbing it as he would typically grab a female.

With so little elapsed time, two FAPs come into play. The mating male still experiences amplexus, effectively tying him to the female. The male that grabs the clutch also may have entered a fixed action pattern, although in his case, he grabs the egg mass rather than the female.[24] Caught in the throes of this FAP, the egg marauder forces open the egg case with his hands and ejaculates into it.

These secondary males effectively hijack the clutch by releasing their sperm into it. This manoeuvre is known as clutch piracy. Sperm from

---

[24] In other cases, a male may attempt fertilization hours after the female lays her eggs. In these cases, the clutch rather than the female must trigger the behaviour.

the pirates enters the fray after the sperm of the mating male. Many sperms from the first male have already died in their attempt to break through the protective coating of the eggs. Sperm from the new male can now outnumber the sperm from the first male. Where fertilization has not yet occurred, the hijacking male has a better chance of fertilizing the eggs than the actual mate.

Clutch pirates also have a second line of attack. Despite the best efforts of the sperm from the original male, it can usually only reach the eggs on the outside of the clutch. Pirates will tear a clutch open, potentially damaging the outer eggs to expose the unfertilized eggs at the center of the clutch. The fertilization rates for the pirates can exceed the fertilization rates for the mated males.

Even pirates have competition. DNA studies show that as many as four different males can father a clutch (314). All but one of these fathers has bypassed the choice of the female. All of these males have mated despite the labours of the male who claimed the female as his prize, thus sabotaging his reproductive efforts. These males are the very definition of sneaks.

Many sneaks face the similar dual problem of getting past a jealous, mate-guarding male and finding a way to mate with an unimpressed female. Historically, we have considered such males to be subdominant or less equipped to win at the mating game. Now, we realize that sneakiness can sometimes outwit dominance and brawn.

The side-blotched lizard lives in western North America from Mexico to southern Canada. Strong dominant males have orange throat markings. They will stake territory and defend it against less able males. Males who are incapable of maintaining a territory on their own choose a different strategy. Small groups of these blue-throated males cooperate to hold smaller territories and thus gain access to some females (317). These males have at least one advantage over the sexier dominant male: their lower levels of testosterone give them better immune function and lower susceptibility to parasites. That is, they live longer and may thus have a more extended period to mate, even if they have fewer mates per unit of time (318).

Unfortunately for these social males, fewer females visit their territory than the territory held by the more powerful macho males. Because of their higher mating success, one would expect that the genes for the larger males would become more common and that cooperation, as a strategy should die out. However, behaviourists have watched these males for many generations. The proportion of these populations wobbles, but both the orange and social blue-throated lizards remain in about the same proportion over time. Something else must be happening, and that "something" is a sneaky third player in the mating game.

The side-blotched lizard has three, not two, different types of males. The third player is a yellow-throated male, one that mimics a female. The female mimic invades the territory of the dominant orange-throated males. He waits for a real female to make her debut and then presents himself to the territory holder as a potential mate. He then mates with the female right under the nose of the territorial owner.

The strategy has high risks and payoffs, and the success of this usurping strategy can vary (319). The sneaky female-mimic lowers the fitness (offspring produced) of the aggressive loner males. This offset in the number of offspring produced by the loner suffices to permit the cooperative strategy of the blue-throats to succeed (317).

One other factor plays a role: female choice. Rare traits can attract the females of many species. Female side-blotched lizards share this penchant. When one of these three morphotypes becomes less frequent in the population, the females become more attracted to that kind of male (320).

Together, the three forms of this lizard—aggressive, cooperative, and sneaky—and the preference of females for rare males, form a stable community, with no male strategy having an advantage over the other.

Lest you believe that lizard sex is just weird, we have also found this tripartite mating strategy in other species too. Dominant male ruffs, a type of sandpiper (bird), will recruit males with less spectacular plumage as "satellites." The presence of the extra males attracts females. Satellites will help defend his mating territory in exchange for the occasional chance to mate as females come to him. A third, genetically determined type of male, a "faeder," imitates a female (321). The faeder casually associates with the ladies. Eventually, one of the females will bend her head in a mating posture. The faeder then quickly moves between her and the dominant male and copulates (322).

A fish known as the midshipman (genus *Porichthys*) employs a slightly different sneaker strategy. Their early childhood conditions determine if they will be large males or small males. The larger fish can attract mates much like songbirds and frogs. They make a sound that resonates through the water. Females will follow these melodies until they come upon the fish of their dreams.

If small males play fair, then they don't stand a chance in these competitions. Males that get stuck with small bodies are not necessarily small in all aspects. They can have testes that comprise up to 15% of their body weight. The smaller fish will wait for a romantic moment between an amorous couple and then sneak into the fray. The romantic male may mistake him for another female. Because he produces far more sperm than the larger male, the sneaky small midshipman has an excellent chance of pirating part of the clutch of eggs for his genes.

In the stickleback fish, a male must make a nest and then entice females to it. Nesting presents more energetic challenges to small males relative to large males. Because of their small size, they cannot make a nest that will attract females. These weaklings will sometimes sneak into a nest of a competitor *before* the female's arrival, while the nest owner busies himself with courtship. The pirate leaves behind his sperm. This pre-emptive move allows him to fertilize a few of her eggs when the female lays them (323; 324; 325).

One species of stickleback goes beyond simple nest piracy. Whenever he sees another male courting a female with eggs, the sneak will stop what he is doing and feign submissive female behaviours. He will skulk along the bottom, looking like a female, and he will wait. When the courting male induces the real female to enter his nest, then the sneak dashes into the nest as well. To the nest-builder, it looks like nothing more than getting two clutches of eggs at once. The sneak, however, deposits his sperm before the owner even has a chance to enter the nest, much less begin mating.

Some of these female-impersonators continue the charade, allowing the nest-owner to mate with him over the eggs of a female. The sperm of the impressionist thus mingles with the sperm of the nest-owner. Hence, the sneak gains even greater access to the eggs (325).

Nature has also designed humans to engage in clutch piracy. We are social organisms, have internal fertilization, and small clutches (usually one at a time), so our males sneak differently. We do not worry about males that can mimic females, but we do express FAPs that help us sneak and avoid becoming the victims of sneaks. Our physiology still programs human males to prevent the sperm of other males from usurping their reproductive efforts.

During ejaculation, the human penis and ejaculum push away any other sperm that might already be near the cervix of the female. Pre-existing and competing sperm accumulates on the top of the head of the penis, which effectively scoops away much of it from the cervix and the female.

Men who believe themselves to be cuckolded self-report that they thrust harder and deeper during copulation than they would typically thrust (326; 327; 328). This increased force occurs when they think their partner has had a sexual encounter (329; 330). It may be an innate behaviour that occurs even when their partner is raped. Such behaviour intensifies the scooping effect and helps the male partner combat the sperm of sneaks.

This piracy function would only have a benefit if the female had two sex acts within minutes or hours of each other. Today, the first sex act might occur during group sex, gang rape, prostitution, promiscuity, or voluntary extra-pair copulation. The second encounter would happen when the resident male insists on sex in response to suspected infidelity (326). If we look to our two most closely aligned cousins, the chimpanzees and the bonobos (a.k.a. pigmy chimps), we can get a glimpse of how this male response to human egg-piracy may have functioned in our distant past.

Male chimpanzees and bonobos live in groups of closely related males. Chimpanzee males are bigger and stronger than female chimps, and they use that strength to dominate them. Chimpanzee brotherhoods form coalitions that harass females and prevent them from forming alliances. The group has both physical and sexual control over the females. The dominant male has the greatest access to females during fertility, but no one-on-one pair-bonding occurs.

Our other first cousin, the bonobos, have males and females that approach each other in size. They are highly sexual, and they can use sex as a medium of exchange. They also live in bands of brothers and cousins, but these males do not dominate the females. Like humans, female bonobos have a hidden fertility period, and they are sexually responsive to multiple males throughout their fertility cycle. Bonobos spend time engaging in copulations in every conceivable position and of every conceivable type. Like us, they have expanded the functions of sex far beyond that of reproduction.

In both of our species, the result is dramatic: to reproduce successfully, a male human or bonobo retains a constant interest in females. In most other apes, males have a heightened interest in a female only when she shows signs of ovulation. Bonobos, on the other hand, have a life spiced by both violence and short copulations, most lasting only about 13 seconds (15).

If our last common ancestor was anything like the bonobo, then we too were once really sexual critters. Our ancestors would have had more frequent copulation with more partners than we do today. The evolution of a ridge on the penis would have given males with pronounced ridges the advantage of a sexual pirate. With any given intercourse, the ridged pirates increase their chance of reproducing and decrease the chances of their sexual rivals. Hence, their genetic contribution to the next generation increased relative to the genes of their competitors.

Humans share the sperm-scooper design with several species, including some damselflies and dragonflies. When these insects mate, they insert a small shovel at the end of their abdomen to remove any sperm that is already in the female (331; 332). Several other families of insects share this house-cleaning ability, as do some crayfish (333), cuttlefish (334), and octopus (335).

Nature loves efficiency. A good design can show up repeatedly across a range of phyla. In the case of all these sperm-scoopers, the females can have multiple partners, fertilization is internal, and males have little control over female sexual expression. Not everyone who meets these parameters has scoops. Other mechanisms for sperm competition and sneak-prevention can also work.

Chapter 8

Mate Guarding

"La donna è mobile (Woman is fickle)"
~ *Rigalletto* by Verdi and Piave

Theft Avoidance and Green-Eyed Monsters

As a species, we take a dim view of sexual cheating by our spouses. Human males and females guard their mates, but on average, males tend to be stronger mate-guarders than females and females tend to guard their resources with greater jealousy. If we again look at the roots of our species—the proto-humans of the Pleistocene era—we see why. Males guard access to eggs. Females guard access to resources for her eggs. The rare egg, not the common sperm, limits access to reproduction. For women, the correlation between fidelity and genetic success is tenuous.

On the other hand, if a male, can gain access to more eggs, he has a better chance of passing on his genes. By copulating more often, or by pirating an egg through removing another man's sperm (clutch piracy), a man improves his chances for reproductive success. If a woman can gain access to resources for her children, then her chances for reproductive success increase. In this system of sexual-selection feedback, the more signs of fertility that a woman can present, the more the signals of success she can demand from a potential mate (336). Women are skilled at picking up indicators of wealth. The subtle signals of old, conserved wealth can thus trump ostentatious displays of spending that can leave *nouveaux riches* again impoverished and unable to provide long-term resources for her offspring (337; 338).[25]

If clutch piracy is a form of theft, so is absconding with food. Parasites are the blatant offenders or resource thieves, but even organisms that humans do not define as parasites can engage in resource robbery. For non-social or marginally social animals, the possessions hidden by one animal become fodder for the next one. Theft is a common crime among animals, and they have adaptations that reduce its impact. Many birds and rodents, for example, cache food. Many will prefer several small, scattered stashes to a few big ones focused in one area. A single large store of food may be easier to guard, but it also constitutes a higher risk. By scattering their food, they make it harder for dirty rotten thieves to get all their hard-earned food (339).

If a western scrub jay knows that another bird observed it hide food, it will return later and move the stash. It behaves as if it knows the larcenous intent of the observing bird (194). It can also make foraging alliances with birds that have proven trustworthy (340). These

[25] I know someone who once gave his girlfriend an expensive designer purse for her birthday. She loved it—until he told her he found it in a resale shop. He saw his action as thrifty. She saw him as offensive and cheap. Hint to everyone out there: DO NOT tell your girlfriend you got her gift at a resale shop unless you really know how she will respond. Usually, it's not pretty.

behaviours tell us that these birds have a concept that is similar to our idea of a dirty rotten thief.

In studies with lemurs, these primates know if a human trainer will cooperate in the finding food or if the trainer will compete with it for food. If the trainer competes, the lemur will use strategies to mislead the scoundrel (341). Primates will also use these tactics in the wild, sending false signals to competitors. Like humans, other primates also use such sneaky tactics to gain access to mates and resources (226; 342).

For individuals who hold subordinate positions within a social hierarchy, deceptions, such as concealment and distraction, become crucial tools of survival (343). One finds deceptive behaviours, like those found among primates, mirrored in birds, insects, and even in plants. When honesty carries a high price tag and lying carries a positive genetic payoff in survival and reproduction, then lying prevails.

The same is true for sexual theft. Among species that form social bonds through mating, a sexual interloper will try to avoid confrontation with the mate. Many species engage in mate guarding. In mate-guarding species, an animal will chastise anyone who tries to take its sexual partner. Most typically, mate-guarding is male behaviour, though there are species in which females will also guard a selected mate. Paradoxically, the guarders may simultaneously attempt to have sex with more than one partner. The philandering male runs a genetic risk. The social mate may also cuckold him, leading to competition with the sperm of another animal (226; 344; 345). However, she may find a way to have an affair even if he remains loyal. Unless the female mate can impose a high penalty for philandering, the male may leave behind more offspring if he has affairs on the side. The female, on the other hand, will not increase the number of her eggs by having affairs with other males.

For the female, such extra-pair copulations, however, can increase the genetic variability in her offspring (346). To gain some desirable trait, she may find another male superior to her social partner (347; 345; 348). In one study of black-capped chickadees (349), those females that two-timed their mates always did so with higher-ranking birds. Through these extra-curricular matings, she reduces the risk of having an infertile mate sire all of her eggs; she simultaneously increases the quality of genes to pass to her offspring.

Mate quality does not always show this same, straight-forward relationship with cheating. Female swallows, like the chickadees, also cheated less on desirable, highly ornamented mates than on lower-quality mates (350). However, these female birds show subtle physical (phenotypic) traits that correlate with their tendency to philander. It thus remains unclear if the lower levels of cheating are due to the quality of the male or an initial choice of non-cheaters by the males.

How males prevent philandering also impacts the tactics of female sneaks and the likelihood of success for the swinging females (351). For example, males can watch their mate, drive away rivals, or monitor the sexy communications of the surrounding males (352).

For many species, it is the size or quality of a territory that determines copulation, not other sexy characteristics of a male. In many species, males hold territories that attract females. Females choose a territory and mate with its owner. In other species, females also hold territories. If the territory of the female overlaps with the territory of more than one male, be she a lizard or a fox, she then has more than one mating choice (353; 354). Males, therefore, may profit from recognizing their neighbours and by expressing aggression toward them (355).

Dragonflies and damselflies rank among the best mate guards in the world. The forms of mate-guarding vary with the needs of the species, and some dragonflies defend their sexual asset more effectively than others. A male and female may remain in direct contact, or the male may simply keep watch over her as she lays her eggs. In contact-guarders, the couple will fly together, with the male firmly copulating with the female. They can remain in this position for hours. Contact-guarding occurs most frequently in species that must distribute their eggs in several places (356).

The reason for the Kama-Sutra-style joining is simple. Male dragonflies have that specialized sexual organ that acts like a scoop (357). If another male gets hold of the female, he will scoop away the sperm from the first male and deposit his own. For the first male, the story ends in a genetic tragedy: no sperm, no fertilization, and no continuation in the gene pool.

In other species, the mate will copulate, then continue to stand guard until the female has laid her eggs in his mating territory (358). This post-copulatory guarding happens in species with rare females. It allows the male to increase his success because guarding allows her to lay eggs up to four times longer without interruption from another male. The act of mating seems to trigger the guarding behaviour, a fixed action pattern. After mating, males will even guard other females in their territory with whom they have not mated (332).

Satellite Males

"Later, when the anger of King Ahasuerus had subsided, he remembered Vashti and what she had done and what he had decreed about her. Then the king's attendants proposed, "Let a search be made for beautiful young virgins for the king. Let the king appoint commissioners in every province of his realm to bring all these beautiful girls into the harem at the citadel of Susa. Let them be placed under the care of Hegai, the king's eunuch, who is in charge of the women; and let beauty treatments be given to them. Let the girl who pleases the king be queen instead of Vashti." This advice appealed to the king, and he followed it."

~ Esther 2:1-4, *The Bible,* New International Version

By the standards of most animals, King Ahasuerus behaves like a typical male. He is a dominant primate who can use his power to gain access to females of his species. Ahasuerus commands his inspectors to find beautiful young virgins. The meaning of his command translates as "find the fertile females who are not impregnated by any other male." Because of his social stature, Ahasuerus can gather a harem too big for one man to control.

To tackle this problem, Ahasuerus needs to enlist the help of other males. Biologists call these partners who help defend the territory of a dominant male "satellite males." Ahasuerus, like the males of many species, did not trust the satellites. He knew that they would be more than willing to inseminate his females if they had half a chance. Consequently, he used satellites whom he could rest assured would not become the fathers of children of the harem, even if sexual intercourse were to occur. He used eunuchs.

Other animals cannot easily make eunuchs on demand. Otherwise, they do not differ from us very much, and they come to similar solutions to the problems of mate control. A male cannot always maintain a territory on his own. Sometimes, he needs to recruit help. In species that have similarly sized males and females, a male who holds a territory often prefers females as co-owners than other males. If they need to garner the help of another male, the owner may prefer a close male relative, i.e. someone who shares many of his genes. Similarly, males prefer their male relatives as neighbours (359; 360).

Territorial males know that if they allow another male into their territory, then that satellite male may find an opportunity to attract local females. The intruder will also have access to food and other resources within his territory. Allowing access, and to whom that access should be allowed, thus become significant life decisions for a territorial organism.

In territorial species, a male uses a cost-benefit approach to the issue of territorial defence. For every bit of territory that he controls, the male pays a price. He must patrol it, watch for intruders, and drive intruders away (361). It takes energy to maintain ownership. He expends energy defending the territory, marking it, and keeping it secure. He acts to attain the goods that he needs, especially resources to obtain the all-important egg from the female.

In good times, a well-adapted male may have little trouble defending his prime piece of real estate against intruders. If the number of intruders remains low, and if the intruders are not too desperate for admission, then the cost of the territory remains low. However, when his health falters, or when competition for territory becomes too intense, a dominant male faces a dilemma. In times of famine, drought, or overpopulation, limited territories may be under constant assault by non-landed members of his kind. He may be forced to cooperate with other males to defend his territory successfully.

It may make sense for him to risk a bit of infidelity in exchange for security. If he can get other males to cooperate with him for their mutual benefit, then both he and the cooperating males reap the rewards of the alliance. It is akin to paying the cost of a security alarm system. The price takes away from the overall currency (offspring) of the owner, but it can be very worthwhile in the end (362).

If your brother happens to run the security company, then all the better. If they are sneaky enough, then cooperating males gain access to the resources like females—the currency of the wild—but the owner will keep the bulk of the wealth (gene transfer) for himself. If your brother is your conspirator, he may get access to the females, but he will also pass some of your genes to his offspring.

In many species, the satellites act like soldiers of fortune. They offer their services as agents of protection in exchange for access to a portion of the territory controlled by the owner. Often, they will patrol the frontiers of the range, chasing away other males and risking injury to retain their satellite status. The owner may lose a few eggs to the sperm of the sneaky satellites. In exchange for this concession, he may gain greater longevity and health, leading to more babies in the long term. In these cases, the subordinate males may sneak copulations when away from the watchful eye of the owner; the positive impact of their cooperation for the owner, however, would outstrip this cost of doing business (277).

Among water bison, for example, a dominant male tolerates subordinates in his territory, despite these disadvantages, because the satellites will help him defend his territory and his herd in exchange for a place to live—and possibly sly access to females (363). These satellites also have a better chance of taking over the territory than males outside the herd.

The same principle works in fish. Male guppies are obsessed with sex (262). If there are just a few males in an area, the more aggressive males will try to chase away the less fit males. Even the most aggressive male can only handle so many competitors. Aggression usually gives way to competition if a river (or a fish tank) is crowded. The purpose of the aggression is, of course, to monopolize females. The displaced males will get as close as they can to the courting male and his female. If the courting male becomes distracted, a satellite will seize his chance. He will approach the female, follow through with a quick courtship dance, and quickly inseminate the female guppy.

Males sometimes increase their reproductive prowess if they live in a "harem" of satellite males. In one African cichlid fish, *Pelvicachromis pulcher,* the cost of defence per sired offspring decreases quickly when they live in such harems. The males who can tolerate satellite helpers will reduce the energy needed to defend their harem and offspring. Consequently, tolerant territorial males increase their reproductive success by cooperating with their dependant helpers. Together, they fight intruders who would displace them for access to their females (364).

Satellites receive a clear benefit should a dominant male accept them. The territorial owners also benefit in ways beyond the mutual defence pack. The associated males can function much like a human "wingman," who helps a buddy get a date. When satellites sing, display, or court females in other ways, they can act as an extension of the dominant male. Their extra dose of posturing and swagger may attract more females to the dominant male. His control of the subordinates may enhance his status with the ladies. Satellites benefit because they can sometimes surreptitiously intercept the females en route.

In many primates, sexual aggression by males, including subordinate males, gets rewarded with a life-long increase in progeny (365; 366). In hamadryas baboons, low-ranking males have little opportunity to mate. The head of the troop runs the harem. He is all-powerful, and usually not willing to tolerate competition. One can view the lower ranking males as satellites. Like most sexual satellites, they can obtain access to the females by using sneaky sex. Males guard their females, but the would-be thieves closely monitor the behaviour and position of the alpha to seize their chance (367). Cooperation with the alpha has benefits, but so does betrayal (368).

To gain access, the stealthy lower-ranking baboon must first find a good location. Usually, he will engage in sex behind some visual blind or near a good escape route. He will court sexually active females when the dominant male is otherwise occupied. In baboons, this courtship may include grooming.

The female hamadryas knows the rules. The alpha male will punish her if she engages with the lower ranking male. Like these males, she also knows how to cheat when it suits her needs. If she chooses to groom him—a behaviour that acts like human consent and foreplay—she will try to do so in private. If she cannot slip away for a grooming trist, she will sit far from the alpha, with her back toward him, and with the sneaky male in front of her, hidden by her body. Should they be discovered, the alpha male would attack.

For the subdominant male, courtship has short-term rewards that go beyond sex. With his lower status comes less access to grooming. Without grooming, his skin can accumulate parasites, such as ticks and biting insects. The presence of parasites annoys and weakens the male, and the pain and annoyance contribute to his continued low status. The courtship grooming removes parasites from his fur, an act that can have long-term consequences for both status and reproductive potential. If caught, the alpha severely punishes the offending females and their low-ranking consorts.

Among the hamadryas, dominant males have another means of controlling access to females. An alpha will harass the females under his control. He chases them and lashes out for minor infractions like an abusive spouse. If she wanders too far from the alpha, he will punish her, usually with a bite (369). Females fear the alpha. This fear may make them less willing to risk a dangerous violation, such as unapproved hanky-panky with another male.

The hamadryads have an extreme system of social organization. Other baboon species protect the genetic investment in different ways. The male yellow baboons (*Papio cynocephalus*), for example, will mount fertile females in order of rank. The dominant male gets priority access during ovulation, giving him the best chance of fertilization (370). Unlike the hamadryads, dominant yellow baboons usually control access to only a few females.

Dominant male chimpanzees, in contrast, will sexually harass females even when they are not sexually receptive. This ongoing sexual aggression correlates with their chances for paternity (371). It is part of an overarching pattern of sexual dominance and mate-guarding. This aggression may also give the dominant males a reproductive edge by increasing his chance of becoming the first to inseminate a female during the early portion of the fertility cycle.

From the perspective of the female, having sperm from an alpha male is a good deal for her offspring. Because her agenda coincides with the desires of the alpha, her behaviour reinforces his chances of fathering her offspring.

At the end of copulation, the yellow female baboon also emits a copulation call. This vocalization encodes information about the perceived quality of her mate (372). When lower ranking males hear her call, they understand who just had sex with her. If the call says the male has a high rank, they approach with more caution, giving the sperm a better chance of reaching the egg. On the other hand, if the call says she has just copulated with a low-ranking male, the higher ranking baboons will quickly approach to chase away the dolt and to mate with her themselves. By calling, the female has a better chance of accessing quality sperm than if she were to remain silent.

The alphas of these two species of baboon use strikingly different methods to protect their reproductive monopolies. Dominant hamadryads chase low-ranking males from the females, bully the females into submission, and keep a vigil to protect their sexual territory. The dominant yellow alphas share access to the females, but only at their pleasure. They have a right to choice copulations with females, but lower-ranking males do get a chance to have sex and potentially to reproduce.

In both of these species of baboons (*Papio*), males defend their access to females against interlopers and low-ranking males. The strategic difference comes from different pressures in their environment. The hamadryad baboons live in open arid regions. There, the alpha can easily watch and harass more females than can the yellow baboons. Because of their landscape, grooming and copulation with other males will get the attention of the dominant hamadryas, and he can then take action to stop it.

In contrast, the yellow baboon lives at the border between the savannahs and forests. This simple environmental difference demands different deterrents in the two species. One environment permits visual cues, and the other environment does not. The troop depend more heavily on vocal communication among its members. A dominant male yellow cannot control as many females because he cannot see them through the vegetation. This simple difference has led to different types of societies, different mechanisms of control, and different ways of cheating the system.

A Resource is a Resource is a Resource

Pied wagtails, a riverine bird, also use satellites, except that the resource they need to defend is food. During the winter, the birds need to get into shape for the breeding season. When times are good, a bird will not waste its time or energy defending territory. In winter, when the hunt for food becomes harder and the take becomes leaner, a healthy, bird will become territorial, guarding its precious resource.

The more able birds control the choice feeding territories near rivers and streams (373). Birds that cannot defend a territory will retreat further inland. There they will scrounge for scraps. When hunger, a potent force, overcomes these lower ranking birds, they move toward the occupied territories along the river in search of food. These outcasts have little to lose. If they do not find enough food, they will die without reproducing and be out of the gene game. During a famine, when they will move to oust the masters of the claimed feeding territories, either they will find food there, or they will die.

The owners do not like the intrusions. In contrast to the intruder, the owners have everything to lose. If they let the new birds into their feeding territory, these newbies will have access to their hard-won caloric resources—food. If they fight every desperate bird that enters their territory, they will exhaust themselves. At that point, the owner becomes easy pickings for others of its species and for predators.

The wagtail owners are good historians as well as good economists. In addition to their cost-benefit analyses, they remember an old Roman adage: *The enemy of my enemy is my ally.* To help them defend their besieged territories, they engage in conditional mutualism. They will tolerate a limited number of intruders into their territories in exchange for services. First, the owner allows in females, then weaker males (373). In exchange for an optimal food source, the tenants help the owner defend its resources. By allowing these birds into his territory, he enters into a "contract" with them. He effectively gives them hunting privileges in exchange for aid in defending the territory.

Despite their tacit agreement, the owner will openly express his displeasure with the situation. Whenever the satellites test the limits of their power with a "chis-ick" call meaning "is anyone home or is it safe to take charge?" the owner responds with an aggressive call that roughly translates as "Watch it you dirty so-and-so. I own this land, and I'm ready to defend it" (374).

Both females and weaker males become satellites to dominant landowners. From the perspective of the owner, accepting a few tenants constitutes a far lower threat than the surrounding owners who may wish to increase their land holdings. He can take advantage of their desperation, trading food for temporary services.

As among humans, the owner has no real intention of honouring the contract. Once he no longer needs them to defend against the powerful neighbours, or if they become too uppity, he can evict weaker satellites from his river territory more easily than he could expel another fit male. Once the owner again becomes capable of defending the territory by himself, he intends to expel the satellites. Although the owners send signals to the satellites that they are tentatively welcome, this welcome comes from necessity and not from the heart. When the satellites cease to be of service, the owners evict them without qualms.

As a species, humans also welcome satellites into our homes, neighbourhoods and villages with the same devious intent as the birds. The people with the resources to do so will hire domestic help like nannies and security guards, but the helpers never become a real part of the family circle. When they no longer serve a useful purpose, we usually dismiss them or fail to provide for them.

In many systems of slavery, including the American version, elderly slaves would receive less strenuous duties, such as feeding the active workers. They also received less clothing and provisions (375). Sometimes, they remained economically useful as mediators and mentors for the younger slaves (376 pp. 5-6). As their economic usefulness became less and less, their care often fell upon the compassion of the slave community as the master reallocated his resources to his more valued workers (his satellites).

Brazil, whose abolition movement started later than the antislavery movement of the United States, adopted a policy of gradual abolition in 1885. Emancipation started with the elderly slaves—those with the lowest economic value— (377). An earlier attempt to free children first failed miserably. Slaves freed in old age typically had no pension and no means of caring for themselves.[26]

Through history, it has been far more common to free older slaves from bondage than to free younger ones. Rome had one of the best emancipation scams (378). Some slaves received a small amount of money across their service. They could save this cash to buy their freedom. In doing so, they gave the money back to their masters, who would then reuse that money to buy a more useful, younger slave. In essence, the slave became a bank the master used to purchase another slave. The wage-theft then continued with a new satellite.

We may welcome security personnel into our homes to hunt and to expel an intruder, but they are not welcome to live in the spare bedroom. When we need help in our factories and farms, we may open our borders, but when we can again cope without the influx, we will just as quickly limit access to our territory. On the surface, we may praise the teachers who educate our children and the firefighters who protect our homes, but in reality, many of us are willing to engage in wage-theft by revoking the pensions of these satellites when it is to our economic benefit.

---

[26] There is an obvious economic advantage to the slave owner for not paying for the upkeep of elderly slaves. Today, a parallel issue has arisen in the current post baby-boom period. It is now commonplace for corporations and governments to underfund negotiated pension plans and to raid these plans to fund other popular projects. Workers who have dedicated their lives to state or corporate service now find themselves victimized by these practices as they reach old age.

Female Mimics

"Girls will be boys, and boys will be girls. It's a mixed up, muddled up, shook up world except for Lola."
~ "Lola" by Ray Davies

We humans have complicated psychological lives and psychosexual lives. It seems to be the price of having a big, active brain. We often flatter ourselves with the idea that our behaviours are unique in the animal realm. Take transvestism, for example. Surely, there is nothing as remarkable in the animal world as our ability to identify with the sexuality and sensuality of another gender. Surely, no other male animal can successfully convince surrounding males that he is indeed a female.

Wrong. Female mimicry occurs in many males. Such mimicry can let a rival male slip by the eyes of a watchful male guard or even a disdainful female. Ethologists—scientists who study animal behaviour—thus classify female mimicry as a sneaky behaviour. Female mimicry can also help an individual to gain access to an otherwise inaccessible feeding territory.

The world of crustaceans includes crabs, lobsters, and many other jointed-legged animals that taste good with garlic butter. It also includes a group of marine creatures known as isopods. The isopods include the roly-polies or woodlice that inhabit many gardens and woods. These land creatures have most of their cousins in the oceans.

One of these marine isopods (*Paracerceis sculpta*) has three body forms. Females find the big alpha males very sexy. They willingly follow such an alpha into his boudoir—a hole in a sponge—and there they mate with him. The intermediate-sized beta males look like females.

Female *Paracerceis sculpta* prefer big, territorial alphas, so the smaller betas have little chance of honestly attracting a sexual partner. Instead, they exploit their similarity to females. Their feminine appearance allows them to approach mating pairs. The mimic even acts like a sexually receptive female (379). The alpha male permits the entry of the mimic into the sponge—probably because he thinks he will gain access to a second female. Instead, the female receives competing sperm from the beta (380).

Paracerceis sculpta males have another sneaky tactic up their exoskeleton. A gamma male is even smaller than beta males. He looks like a small juvenile that cannot yet produce eggs or sperm. He also easily invades the bedroom of the alpha and deposits his sperm there as well.

In one tropical rove beetle (*L. versicolor*), larger males control feeding territories and harems of females. The females help him protect their territory, which is usually a lovely piece of dung that attracts yummy flies. Unless they are very clever, smaller males typically get left in the cold without a feeding territory and mates.

Some of these small rove beetles can pass as females and join the harem. By posing as females, they can remain at the dung site and may even copulate with the real females. They may pose as females their entire time at the dung pile, or they may pretend to be females only temporarily, such as when unable to defend themselves during feeding or courting (381).

The isopod and the rove beetle have no choice about their physical deceptions. Their bodies do not change once they hit maturity. The small animal will always be a small animal. The rove-beetle mimic will always look like a female. Throughout their lives, their deceptive behaviour becomes their only chance of ever mating.

Among the cephalopods, the deception goes beyond a fixed appearance. By the standards of the invertebrates the cephalopods—cuttlefish, octopus, and squid—are Einsteins. Their brains give them more flexibility in their behaviours than the insects have. Cephalopods can adopt a deception and then discard it when it no longer serves its purpose. The bottom-dwelling octopus complements its big brain with extraordinary camouflage. It changes its skin colour and texture to match its background.

The open-water squid are not as adept at the texture-trick, but some have significant control over the appearance of their skin. Many can mimic the animated rippling of the sun across the water, making them less visible as they swim.

Among the cuttlefish, there are real studs that seem to get all the lady cuttlefish. These studs guard their mates zealously. Then there are the nerds that cannot find a date. Technically speaking, we would say that they have very poor fertilization success. Some romantically challenged cuttlefish have found a new use for their colour-changing skin. They have modified the camouflage into a sexually sneaky tactic. They change the colours and patterns on their skin to make themselves look like females (382). This cunning tactic gains quick access to females and leads to nearly immediate fertilization. Translation: the cuttlefish gets laid.

Why would these males be so successful with this stealthy approach? Female cuttles are very fussy about consorting with males. Males will guard their mates and not let other males come near. As with the isopods and the rove beetles, the mimicking males can sidle up to the male and his female, with the goal of copulation still present. This disguise allows him to get past the vigilant eye of the consort male *and* the fussy female (383) who may not even realize the sneaker is mating.

The cuttles are not alone in the world of large-brained sneaky sexual mimics. The blue-gilled sunfish runs a similarly devious racket. Small sunfish females prefer large males. If the male sunfish is small, he does not have a prayer of making an honest hit with the potential mates. Instead, the smaller males, who are about the same size and colouration as the females, will approach a mating pair. He will then get between the large male and his female. The male dupe treats the mimic as if he were another female. The mimic seizes this opportunity and deposits sperm onto the eggs even as the dupe tries to fertilize the sperm of the mimic. Like the dominant male, the sneak has a high fertilization success because he releases his sperm alongside the sperm of the sexy male. This proximity gives him an advantage over those males who play fair and do not try to sneak (384). Sometimes, honesty just doesn't pay.

Sneaky Females

"Money makes the world go around"
~ "The Money Song" from *Cabaret* by John Kander and Fred Ebb

Over the years, my male students have protested vehemently when I discuss the concept of the sneaky male.[27] They protest that women are

---

[27] The sneaky-male theory first developed in a very limited sexual context. It tried to explain why the desire for particular traits in males (such as dominance) did not amplify those traits in a population beyond control. In other words, why does male aggression not get worse with each generation? Why does variety exist among males despite very strong sexual selection by the females? The sneaky male theory answered at least part of that question. Male variation continued to occur because some males

just as down and dirty as men. They resent that we classify so many of the behaviours of males as sneaky, while females seem to get off without any blame.

I agree. Males do lie to get sex. If there were no copulation at the end of the sneaky male behaviour, it would not count as sneaky because it did not circumvent the female choice in sexual partners.

My students, however, are right. Females lie for sex, too. Animal behaviourists initially overlooked this point. The basic reasoning held that females, being the choosier sex, did not need to sneak. This idea, it turns out, is nonsense. Females often need to circumvent the sexual advances of bullying males. Males and females both lie to acquire the necessities of life, including food and access to reproduction. However, we historically *defined* the sneaking phenomenon as a special case regarding those behaviours that increased reproductive success in males.

Females also sneak for the sake of procreation, but they appear to sneak in a different way than males. Males sneak to gain access to their limiting procreative resource: the eggs. Females sneak to gain access to *their* limiting procreative resources, which usually do not include sperm. In most species, males are too willing to donate sperm with no questions asked. Females may, however, sneak to gain access to their choice of sperm.

Males produce much more sperm during their lives than females produce eggs. They can potentially mate many times over, fertilizing a new female each time. For the female, once she has chosen a sperm

---

could copulate through deception despite the preferences of females and the interference of other males.

donor, she often is stuck until she can lay those eggs. Even if she can mate more than once, it takes energy to make eggs. It takes far more energy than it does to make sperm. An unfortunate choice of mate costs her dearly.

For these reasons, females tend to be much choosier about sexual pairing than males. In most species, nature has hardwired female animals to seek the best situation for their resulting offspring. Nature hardwired male animals to find the most mates. Each gender directs its strategy toward leaving the maximum number of genes in the next generation. For males, that genetic success often means many offspring with many females. In ideal circumstances, he can do whatever it takes, including sneaking, to leave his offspring behind. For females, who are more limited in what they can do, a good strategy is to leave the best offspring behind in the best circumstances for their survival and reproduction.

Some females take resource-sneaking to the extreme. When a female praying mantis accepts the courtship behaviour of a male mantid, she has a hidden agenda. The male has to prove himself a worthy mate by courting her appropriately. He also has to be careful. If she is both willing to copulate with him and a bit peckish, she may also see him as a protein source to nurture her eggs and their children.[28] In many species, a female mantid willingly turns the courting male into dinner.

To help him overcome the handicap of head loss, the male mantis has evolved a very clever trick: zombie sex. His body will continue copulating despite the inconvenience of decapitation (385). The

---

[28] Contrary to the popular image of mantids, females do not always eat their mates. In nature, this cannibalism happens only when the female is feeling a little peckish. For that reason, male mantises seem to prefer fat females that are well fed to svelte females (649).

decapitation even helps him copulate with greater vigour (386). Even if he has not yet begun copulation, his body will continue to search for the vagina of the female, permitting him to inseminate her even after losing his head. Just when you thought that you had seen every possible weird behaviour, zombie sex in mantids reaffirms one's faith in the ability of nature to astound us.

Sneakiness and Resource Allocation

"Diamonds are a girls' best friend."
~ Lorelei Lee, in *Gentlemen Prefer Blonds* directed by Howard Hawks

In many birds, females will choose a good provider-male for a social mate but then sneak away to mate on the sly with other males. When we check the nests of "monogamous" birds for paternity, we often find that more than one couple has contributed to egg production. By gathering what she perceives as superior sperm, she gives herself an edge: higher hatch rates of healthy offspring (387). Females of many species regularly engage in polyandry or mating with many males. She may mate with one male, accept his resources and his help with the nest, but then copulate with the sexy bird in the next territory whenever she gets a chance. Social sneaking accompanies her amorous advances.

For the female bird, having the resources provided by a male is critical to the upbringing of her eggs. She can derive a great benefit if she has access to the sperm of other males. She may also get food or other resources, such as feeding privileges, from the second male. Also, the more males with whom she can copulate, the more genetic combinations she will have for her offspring.[29]

Females may also abandon one set of offspring if the chance to create a better batch of hatchlings comes along. In some birds, sneaky females hide their eggs from their mates. In this way, the mothers can hedge their bets, leaving open the opportunity to abandon the nest before the fathers know she has laid the eggs. This occasional behaviour appears when a better gene donor becomes available. With the eggs hidden, the mothers are free to move into new nesting grounds with their "Mr. Goodgenes" [with reference to *Looking for Mr. Goodbar* (388).]

---

[29] It is not in the best interest of her cuckolded mate, of course. He must use his time and effort to raise another male's genes.

Chapter 9

Lying Plants

"Even Boston beans do it"
~ "Let's Do It" by Cole Porter

Flower Power! Sex and the Single Flower!

~ Allen Ginsberg and paraphrased from Helen Gurley Brown

Virtually every living thing lies. Animals, bacteria, fungi, and even viruses are all liars that cannot be trusted. Organisms can believe each other only if they have complementary agendas. Be it bumblebees or flowers, organisms will practice deceit if and when it gives them an advantage over their competitors, and they will cooperate when it works to their mutual advantage.

So far, we have maligned animals and insects by treating them as if they were the only organisms capable of lying. They are not. We notice their deceit more easily because we know and use so many of the same tricks ourselves, and because we have taken time to observe animal behaviour. Few people watch begonia behaviour with the same intensity as they watch dogs or even spiders, but as I have said before, every living thing lies. Organisms practice deceit if it gives them an advantage over their neighbours, in other words, if it provides adaptive value. Animals mimic themselves, inanimate objects and even plants. Plants can be just as deceptive as animals. They play the same games of mimicry, and they obey the same rules of diminishing returns. Too many mimics and the deceptions begin to fail (389).

In her book, *Sex and the Single Girl*, Helen Gurley Brown (390) broke social taboos about discussing sexuality in young women and busted the myth that every female must marry. She advised young unmarried women to take a critical look at their options and to understand how each potential suitor could harm or benefit them. Brown told her audience where they might look for sex and what styles of sex might be available to women. She showed how appearances—of one's self and one's home—could influence the behaviour of a potential sexual partner. Ms. Brown even advised how to end an affair. The smash-hit book portrayed a romanticized image of a sexually liberated modern woman (390). Brown helped to instil a sense of sexual power in North American women.

Brown neglected to mention that blossoms got there first. Flowers physically select their potential pollinators, and they sometimes decide to use them for purposes other than sex. Like animals, they interact in different ways with different species around them. Styles of sex differ among plants from very independent pollination by the wind to committed relationships with pollinators that have lasted millions of years. Their appearances influence the behaviour of potential sexual partners, even to the point of death for the pollinator. Most plants end their sexual encounters with insects quickly and easily, while others will hang onto the insect long after it wishes to leave. Flower power lives (391).

When we smell the beautiful scent of a flower, we have the unmitigated arrogance to believe that the plant produces that smell for our enjoyment. Blossoms are the sexual organs of plants. They produce that smell for the benefit of the insects they use for pollination, and not for our benefit.

Most of the time, flowers have designed their scents to attract, and sometimes to fool, pollinators. Most insects and plants have developed a mutual understanding. The insect gathers pollen and nectar from the blooms. Flowers produce nectar as a reward for spreading their pollen. They attract insects with bold advertisements of their rewards. In exchange, the insects carry the pollen from spray to spray. The system works well most of the time.

Many plants are scrupulously honest with their insect partners, although a few exaggerate their rewards. Goldenrods call to wasps and bees with their yellow colour. They tell wasps that they have nectar and pollen to spare, but their bragging exaggerates their level of bounty. Goldenrods get away with their claims because these golden-yellow blooms also agree to participate in the warning colouration signals of these insects. Once they perch on these spiky flowers, a few wasps can look more like a swarm. This communication between the plant and the wasp is so good that entomology students, who rank among the few animals that will knowingly pursue social wasps, love to hunt for them among the goldenrods. There, they can find both wasps and many of their mimics.

However, other less scrupulous plants take advantage of this mutualism and subvert it for their selfish ends. Orchids, in particular, are horrible fibbers. About one-third of the species of orchids engage in some form of deception directed toward their pollinators (392). Just as with the animals, deceptive plants must break the code of their pawn.

What works in one situation may not work in another. If its targeted dupe is a sight-hunter, then the plant will use shape, size, and colour to attract it. Prairies, with their vast expanses of open spaces, and the forest floor in spring produce many beautiful and showy flowers because of this visual advantage. If the mark uses smell as its primary cue, then the plant must mimic an attractive odour to convince an insect to come hither.

Depending upon the species of the dupe, the deceit can be the fragrance of a rose or the stench of a rotting carcass. The scent can speak directly of nectar, or it can lie. The flower might signal a communal warning, it may hint that a voluptuous lady-love awaits, or it can warn that a sexual rival has invaded one's territory.

Sometimes conflicts arise over the price of services. Pollinators expect compensation for their efforts. Usually, the payment comes in the form of nectar rich in high-energy sugar. When plants produce nectar, they must devote time and chemical energy to the process. Every drop produced represents energy that the plant no longer has available for its functions, such as growth and the production of blooms, pollen, eggs, and seeds.

A plant that can defraud pollinators and not give the energy-rich payment for its services can then keep that energy—the sugars—for its personal use. Some plants, therefore, try to cheat their pollinators out of their payments. Like humans trying to cheat their employees, wage-theft can be a profitable but dangerous game.

A South African orchid, *D. ferruginea,* can convince a butterfly to pollinate it without coughing up a nectar reward. In the Western-Cape region, this orchid mimics the rich, red, nectar-producing flowers of an iris (*Tritoniopsis triticea)* that grows in its same region. Miles away, in the Langeberg Mountains, the same species mimics the torch lily (*Kniphofia uvaria*), a member of the aloe family that has an orange bloom. The colour match (reflectance spectra) for each mimic is a dead-ringer for its respective model (393). Both models produce a generous reward of nectar for their insect partners. *D. Ferruginea* gives them squat.

In the case of these orchids, human eyes clearly see the mimicry. The colours of the flowers fall well within our visual range. Although the deception is designed for a very different set of eyes, enough of the false signals fall within our visual range to make us aware of the plant's intention to commit theft of labour.

Sometimes the mimicry that fools the dupes may not be quite as noticeable to us. The eyes and brains of the pollinator differ from our own. The intended target may be more attuned to different aspects of colours, ratios of colours, or chromatic variation rather than slavish colour mimicry (394). In one case, a nectar-rich bellflower (*Campanula persicifolia*) serves as a model to a stingy orchid (*Cephalanthera rubra*). The human eye immediately sees the difference between the blossoms, so botanists spent years wondering why bees visited the miser. Eventually, someone did grasp the obvious explanation (395). Bee eyes differ from human eyes.

Every animal has certain perceptual biases (396). Recall that the hamadryad baboons use vision to keep track of female mating, but yellow baboons depend more on vocal signals. In human vision, we fixate on the colour of an object while motion catches the attention of dogs.

Mimicking plants use these perceptual biases to work their sleight of hand. The *Cephalantherid* orchid designs its deception to fool the eyes of the bee. The colour vision of the bee goes much further into the ultraviolet-blue part of the spectrum, where we humans cannot see or see poorly. The bees do not see as well toward the red end, where we see very clearly.

Unlike the red-and-orange-orchid mimic above, this orchid camouflages itself by mimicking the colouration of the model only in the range best seen by its narrow target audience. Once we understood the perspective of the bee, the trick became obvious.

Now recall that African Cape orchid that had red and orange blooms, colours not favoured (or seen) by most bees and wasps. Recall that the spectral *range* measured for the red iris mimic and the orange torch-lily mimic was a very close match for their models, not just the red and orange colours. These deceptive orchids appear to be red or orange to us. The bees will perceive them through their own perceptual bias.

Bee-vision reveals other aspects of the nearly identical spectral range between these flowers. To these plants, what matters is the perceptual world of their insect pollinator. For us non-pollinators, the apparent visual match within our range may serve other yet unknown functions, it may attract other pollinators such as beetles, or it may merely be a pleasing side effect of the plant's attempts to fool someone else.

Some plants take the visual deception a step further. Rather than investing in lots of expensive pigments, they use optical tricks to scatter light. The light-scatter can make a blossom look like it has far more of a colour, such as blue or UV, than it truly has (397).

Because of our perceptual bias, humans assume that visual signals can lure every organism. However, many insects live in a world in which scent can trump vision. Beautiful green bottle flies, for example, can smell a dead animal or the hamburgers at your barbeque from over a mile away.

In addition to misleading visual billboards, many plants produce chemical lures that appeal to the sense of smell. The perfumes of a flower can imitate the sex pheromones of insects (398; 399). Scent has a distinct advantage over sight as a signalling system. Trees, shrubs, and other vegetation can limit a line of sight. The size and shape of the lens of the eye place physical limits on the ability of an animal to focus over long distances. A scent, on the other hand, can travel long distances via the wind.

Even humans, with our feeble sense of smell, key into this kind of signalling system when we look for food. We can smell fresh hot popcorn, strawberries, spicy curries, and other fragrant foods from one or more rooms away. Many potential pollinators have a sense of smell far more sensitive than our own. Some olfactory cues can carry for miles.

Plants release sexual scents to attract male dupes because, among most species, males are usually the ones on the sexual prowl. Males find these fragrances irresistibly sexy. Hundreds of species of orchids lure their pollinators this way, with a promise of sex rather than a reward of nectar. (400; 401; 402).

The scents of these flowers trigger a pattern of instinctive, fixed-mating behaviours in insects. Some of these plants also can downplay their colouration for maximum effect. Many sexually deceptive orchids will cloak themselves in shades of red, green or white—colours that many insects, including their dupes, cannot see clearly (39). By blending into the background, they have less risk of alerting the marks to their presence.

The fooled insects can sense a sexy floral perfume that smells like a female, but they do not see the source of the aroma. The insect comes to the plant with an expectation of sex, but instead, he finds a well-camouflaged blossom. Undeterred, the not-too-bright male tries to mate with the flower because the chemical "pheromone" produced by the plant proves irresistible.[30] The impact on their brains is so overpowering that they have difficulty altering their programmed mating behaviour when faced with this strong chemical cue. In short, they mate with the bloom.

It is the mating motion that scatters the pollen. Sometimes, the male insect ejaculates from sexual excitement. The ejaculum makes the pollen stick to the frustrated wasp, helping to spread more effectively the genes of the plant (403). Even if the flower rewards the amorous insect with food, the poor dupe still leaves the encounter having had sex with a partner unable to pass his genes to baby insects.

Remember, plants usually play this trick on male insects because males are far more attracted to sex pheromones than most females. It behooves plants to attract male hosts for several reasons. Males can "mate" with the plant and still have enough sperm to fertilize the next real female that comes along. Oddly enough, many of these female-mimics are less than perfect. Far from being a flaw, these plants use imperfection to their gain. They take advantage of an innate interest in

---

[30] Recall this phenomenon from earlier chapters. The plant triggers a fixed-action pattern. The scent of the female manipulates the brain of the male so he cannot resist the urge to mate. In essence, the flower is a purveyor of insect porn.

sexual novelty that occurs in the males of many species (399; 404).

One might think that this deception is a potentially dangerous game for the orchid. By feigning sex rather than food, it need not provide energy-rich nectar; thus it does conserve its energy resources. However, if it becomes too successful, it could risk damaging the host population by reducing too many real inseminations. Fortunately for the liar, sperm is cheap, and insects are not as dumb as one might think.

The male wasps may initially fall prey to the deception of the flower, but many of them do learn. With experience, the older males in at least some species learn that the orchid is not a female wasp, and they stop responding to the lie. Later in the season, the males will avoid the plants and find real females. If their strategy includes moving away from the offending orchid, the plant will also have the opportunity to spread its pollen further (405). By forcing the males to move, the local female wasps might suffer a potential drop in available mates; however, they may also benefit from the new genetic material carried by other duped males that move into their area. For these wasps, the loss of a few potential mates is not life-threatening. They have time to learn and to reproduce.[31]

For the small Turner wasp (*Zaspilothynnus trilobatus*), the flower plays a slightly different game. Here the deception looks more like a practical joke. The orchid produces a less-than-perfect female mimic

---

[31] The trick also entails less risk than one might think because of a quirk in the genetics of the pollinating wasps. The females of some duped species can create "haploid" males by parthenogenesis. These males have half the chromosomes of the female. To produce them, the female does not need to mate. Even if deprived of sperm, this species can still produce males. A reduced rate of fertilization among the wasps would actually result in the birth of more males (potential pollinators), which would be good for the orchids (38). In short, the deception gives the flower the added benefit of creating more pollinators in the next generation of wasps. The females also benefit from greater access and competition among the new males.

that releases a pheromone to attract males (406). The males readily respond to the fake female (93% of the time), and they will grab it. That is when they suddenly seem to think, "Wait a minute. . ." Only about six percent of the wasps actually proceed to copulate. The wasps act as if they understand the error and do not ejaculate (406). Instead, they leave the plant and go in search of a new partner. This orchid, *Drakaea glyptodont,* depends upon the ability of the male to recognize the deception and not complete the sex act.

After the deception, the male would rather not repeat his mistake. He will avoid nearby flowers for a while and search for a female far away—on average about 32 meters—from the offending plant. The orchid can thus disperse its genes farther with this sexual ploy than it could with a simple food-reward based pollination agreement. This geographic dispersion results in the pollen reaching more genetically diverse flowering mates. By using sex as a lure, the orchid reduces its risk of breeding with a closely related orchid and producing less viable offspring (407).

Another insect mimic, this time an aster (*Gorteria diffusa*), will produce detailed copies of female bee flies. The details depicted by the plant, including ultra-violet markings outside of human visual range, will trigger fixed-action patterns in their insect dupes. Lustful male flies come to the lure repeatedly (408). Again, the males respond as if the plant was its ideal date.

Before you feel too sorry for the duped bee flies, you should know that they too can be dangerous thieves and liars. Many of these harmless flies benefit humans. Their larvae function as decomposers, as predators on aphids, and as valuable members of aquatic ecosystems. The adults pollinate plants. Others live lives of aggressive mimicry.

Adult bee mimics also lie. Different species imitate honey bees and wasps. One species uses its front legs to simulate the long antennae of potter wasps, and many others move their bellies to emulate the motion of a stinging bee. Some of their best mimics imitate the bumblebees. Unlike many of the other bee flies who simply impersonate another species, *Volucella bombylans,* a particularly good bumblebee mimic, will enter the nests of bumbles to lay their eggs. The larvae of these bumble "want-to-bees" then prey on the larvae of their models, the bees.

Theoretically, little stops plants from using this system of false sights and smells on female pollinators too. The odours would be different. Sexual pheromones drive the desires of males. If this system were to be used with female pollinators, then the plant would need to give the females what their hearts desire. Lady bees and wasps crave resources for their eggs. Those resources could be high-protein prey insects, tasty pollen, or energy-rich nectar.[32]

Some orchids produce inexpensive fake pollen that helps lure pollen-loving insects to the plant (409; 410). Excited pollinators come zooming toward the yellow mass, thinking that they have hit the motherlode. Once they arrive, they usually do find pollen, but the proportions are not as advertised.

Many plants produce this mix of genuine and fake pollens. Others use colour to imitate the colour of pollen. Although goldenrods do produce a large quantity of pollen, their showy yellow colour still amounts to false advertising. Yes, the wasps get a reward, but not as much as they might have hoped (411).

---

[32] Nectar is rich in sugars. Sugars cost the plant valuable calories that it might otherwise put to good use elsewhere. Some orchids mimic other orchids that do produce sugar. The wasp will be unable to tell if she will get a food reward for going to the flower, or if it will just become a pollinator for a stingy, sneaky plant (39).

Other resource mimics try to fool insects that depend upon carrion for a living, including the various flies used by crime-scene analysts to determine the timeframe for the disposal of a corpse. Different types of carrion flies and beetles will invade dead animals at different times during the process of decay. These insects respond to several cues including the size of the body, the heat generated by the bacteria (or other insects) in the decaying mass, and most importantly, the smell. A mother fly will look for these things because they mark the site of resources for her young.

Carrion-mimics develop odours that smell like a good nursery for these insects. The dead-horse arum plant of Italy (*Helicodiceros muscivorus*) develops a stench that explains its name. The sweet smell of rotting flesh that it produces tells female blowflies that the plant is ripe with bacteria, and an ideal place to lay her eggs (412). The arum reinforces its message with a misleading visual cue: an infrared streak that gets warmer than its surroundings, like a tasty rotting carcass (413; 412). The carrion-dependent mothers-to-be find the combination irresistible.

The arum lures the fly into a trap in the flower. There she becomes covered with pollen while she deposits a few of her eggs. Sadly, those eggs will die because the plant is not made of rotting meat, and the bacteria they need to survive do not live there.

Fortunately, the female blowfly is a cautious mother. She withholds some of her eggs and finds another nursery in which she can deposit them for safekeeping. If the arum plant is lucky, the mother blowfly will repeat her mistake. She will move to another arum and thus will deliver the pollen to that second plant. If the blowfly is lucky, she will avoid another encounter with this aggressive mimic, and place her offspring in a proper nursery.

Like the sexy orchids that duped the males, these resource mimics run a risk with this deceit. If brood-site mimics are too frequent, or should the females lay their eggs on too many flowers, then these plants could lose their access to pollinators. If the insect is very common, then the deception causes little risk to the plant. Enough hosts would still breed successfully. However, if the insect is relatively uncommon, the loss of a clutch could be devastating for the plant. When the population of their insect patsies crashes, the plant population would also crash. The remaining flies would then be less likely to encounter one of these cads. The dynamic relationship between the two populations would go through periods of highs and lows, similar to the cycles seen in other predators and their prey.

Sneaky plants have yet another way of using instincts of an insect to their advantage. As with many mammals, many insects display territoriality. When it is ready to breed, a territorial organism will defend its breeding grounds against competition, especially from its primary competitors—its own kind.

In many species of insects, this territoriality is a fixed action pattern (FAP) triggered by simple cues. A plant can use this inborn behaviour to attract a pollinator. Its flower will pretend to be an enemy insect or a competitor of the bug it wants to attract (414; 392). The dupe sees what it thinks is a threat, and it attempts a pre-emptive strike—quite literally. The duped bee will attack the bouquet as it would a real opponent. By imitating the insect, the plant triggers a programmed aggressive response, and the subsequent attack aids in the dispersal of its pollen.

Please Don't Eat Me—Let Me Go

after "Release Me" by Eddie Miller and Robert Yount

Whether they like it or not, plants must share this planet with animals. Animals constantly attack plants for sustenance. With a few exceptions in the deep oceans,[33] one can trace the energy needed to sustain most animal life back to plants. Even the tiger that eats the peacock, or the lion that eats the gazelle depends upon the plants that nourished its herbivorous prey. Those animals that eat higher on the food chain add extra steps to the process, but ultimately, they still depend on the energy produced by plants through photosynthesis.

With the constant assault on their being by animals, plants can be very vulnerable. Grazers—from insects to fish to slugs to mammals—all want to sustain themselves by eating leaves, fruits, bark, stems, shoots, roots, or flowers.

In the face of such an assault, plants have learned how to defend themselves. Poison ivy and hemlock produce toxins that discourage grazing. Others, such as cactus and citrus trees, protect themselves with thorns. The hollow-thorn acacia trees of Central America and Africa have developed an innovative system of mutualism with ants. The hollow-thorn acacia of Central America provides food and housing for an ant colony. In return, the ant protects the acacia from both insect predators and from other plants that want to steal its light. Some species of ants will protect the tree from leaf galls, reproductive

---

[33] The food chains of the deep ocean do not depend upon plants. Instead, they depend upon bacteria that feed on the hydrogen sulfide coming from deep within the earth. Other than the source of the energy, the principle of animal dependency remains virtually unchanged. We still know very little about the ecology of the deep. One can safely predict, however, that some form of deception exists there, too.

structures produced by other insects (415).

An African acacia has a similar arrangement with ants. Although these ants will tend to aphids that feed on the tree (416), they will attack more dangerous marauding grazers, including beetles (417), giant giraffes (418) and elephants (419). These large mammals consider the ants irritating and preferentially eat acacias that do not harbour ant colonies. Some of these plants, like many cacti, have visible thorns. Often, they will also produce honest warning colours of the danger that they harbour, including the presence of ants.

Some plants skip the risk of not harbouring a mutualist ant colony or the costs of keeping them as tenants. The rough cocklebur (*Xanthium trumarium)* has dark specks that resemble ants in both size and shape (420). As noted earlier, many ants bite. Many animals, including large grazers, do not want to cross an ant colony. Ant mimicry can deter herbivores without the metabolic cost of supporting a colony of ants. Even a moment's pause could send the potentially marauding herbivore to another patch of green, thus saving the deceptive cocklebur to reproduce another day. Other plants have body parts that look like aphids or caterpillars. When present, these insects can make a plant less tasty to herbivores (420).

Plants that do not have poisons, thorns, or contracts with ants rely on deviousness to protect themselves. One striking example of defensive deception occurs in *Mimosa*, a genus with many sensitive plants. These plants have a few prickles, but not enough to deter a determined predator. Instead, Mimosas have adapted a standard feature that plants usually employ for a different purpose: leaf movement.

Many plants can move their leaves. They lower them at night by reducing the water pressure, or turgor pressure, in the leaves. A chemical signal tells specific cells in the leaves and stems to move water from a supporting structure, a vacuole, into the surrounding cytoplasm and then out of the cells. The resulting loss of water pressure causes the leaves to lower for the night. The Mimosa performs this same trick in response to touch. The drop in pressure causes the leaves and their supporting structures, the petioles, to collapse.

We do not yet know when or why Mimosa first developed this dishonest defence mechanism. We do know that the trick has at least one deceptive tactical advantage and one real physical advantage. After dropping its leaves, the Mimosa plant looks wilted. A wilted plant does not appeal to a wide range of predators.

Wilted vegetation also deters many bugs that depend upon turgor pressure for feeding. True bugs have mouths that tap directly into the circulatory system of the plants, either the phloem or the xylem. The pressure in these veins pushes the food into the bodies of these bugs. The loss of pressure, even if temporary, would discourage these insects from feeding there or laying eggs there. Other foraging insects, that chow down rather than drink sap, would also look elsewhere for tastier treats. Even mammals may look at a wilted leaf and decide that something must be wrong with it. Thus, the wilting trick helps the sensitive plant avoid predation in two different ways.

Other plants combine a good defence with a good bluff. Many plants shape one part of their body to imitate a more dangerous part of their body. Soft leaves, buds, and fruit may look like thorns. Agave, a desert plant, combines a real defence of thorns with makings on their leaves that look like thorns. The colouration makes the agave far more dangerous looking than it is (421). To be fair, we do not know if their potential predators see the makings as additional thorns, or if they see them as truthful warning signs. An honest warning would benefit the agave because it would be less likely to lose thorns to damage by animals.

By mimicking different animals, plants may attract animals that prey on insects. Queen Ann's lace, a wild member of the carrot family, has flowers with a dark center. This dark spot looks like an infestation of small insects. These lures attract predatory pollinators that want a meat meal rather than a nectar one (422; 423). Like our buddy, the deceived sexy wasp, the little carnivore goes away disappointed.

Okay, You Can Eat Me—Not Really

So far, we have discussed plants that try to avoid being eaten by miming unpalatable plants and animals. Over the last 6000 years or so, since we learned how to raise crops, humans have inadvertently helped many plants to develop another odd form of mimicry: crop mimics. These opportunists have learned to imitate our cultigens.

When we farm, we try to reduce the biodiversity in our fields. Humans have spent much time weeding our gardens and our crops to prevent the land from being overrun with weeds. By eliminating the more visible weeds year after year, we have functioned as a strong selective force favouring less obvious invaders. Crop mimics try to look like a desirable domestic plant favoured by humans. Weeds may mimic the appearance of a crop (vegetative mimicry), the morphology of the seed, the timing of maturation of the crop, or the chemistry of the crop (e.g., tolerance to the same herbicides).

In seed mimicry, the weed has the same seed characteristics as the desired plant. The winnowing of grains to separate the desirable grain from the undesirable weeds has selected for weed-seeds in the same size range as our domesticated crops. Many weeds now have seeds that not only are the same size as their associated grains, but they also have seeds that mature at about the same time.

Until the advent of herbicides, we controlled pesky plants by hand weeding. This weeding became a principle selection force for many weeds because we often missed the weeds that looked like our crops (424) or like our beloved ornamentals. Generation after generation, we removed the more obvious weeds from our crops. This process favoured the survival of those weeds that looked the most like our wheat, flax, or other domesticated plants. Our actions pushed the surviving gene pool toward this mimicry. As mutations arose among the species of weeds, we quickly plucked any whose new appearance made them more visible. If the mutation happened to add to the camouflage, we had an even harder time spotting them.

Weed mimics consider us a predator. Unlike the Batesian mimics among the butterflies, the weed mimic does not try to convince us that they are noxious. They take a very different approach to our predation. They try to convince us that they are friendly and even useful. These *Vavilovian* mimics sell us the story that they are our crops and that we should love and care for them.

Many weeds, such as barnyard grass, have become excellent mimics. We can only recognize them upon close examination of the plant. One flax mimic, *Camelina*, has developed leaves and stems that make it hard to distinguish from real flax. Its seeds ripen at the same time as flax—just in time for the harvest—and its seeds are the same size as the flax seed, making winnowing next to impossible with traditional techniques

(425). We have had to develop more and more elaborate methods to separate this seed from our beloved flax. However, as our sorting has become more precise, we also selected for better *Camelina*. The plant has stayed one step ahead of us, outsmarting us at each innovation.

Our relationships with crop mimics[34] bear similarities to the protection racket run by the cowbirds. Some songbirds fail to recognize the invaders, so they put their energy into raising the offspring of the cowbird rather than their own children. We also have trouble recognizing desirable seeds contaminated with crop mimics. The host birds put effort into raising the cowbirds at the expense of their young, and we raise the weeds to the detriment of our crops.

Like the songbird, we may try to remove the intruder from our field. True, the plant cannot retaliate by swooping down and destroying our crop. However, our millennia-long efforts to control it have led to improved mimics. Just like the songbirds, we now find ourselves spending so much time and energy trying to control the weedy intruder that we often reach a détente with the enemy. As the dupe, we have become willing to accept some losses because we know we cannot eliminate all the parasites.

Occasionally, we find a better way to co-exist. Sometimes the mimics have properties that can make them useful to us. One plant, *Secale cerale*, started its association with us as a mimic of wheat and barley. It snuck into our fields and grew alongside our crops.

We tried to plough this perennial pest under as we tilled the soil in the fall. In doing so, we favoured its transformation into an annual plant with the same growth cycle as our domestic grains. Because it looked

[34] Crop mimicry is also known as Vavilovian mimicry, named for the Russian botanist, Nikolai Vavilov who first described the phenomenon.

like our grains, we often missed it as we hunted for it in the fields. We repeatedly culled it, until only the best mimics remained.

As our wheat and barley grew larger under our domestication, so did *Secale*. Eventually, someone noticed that the hardy Secale seeds were edible and very tasty. We learned that we could grow this less domesticated plant in places where wheat and barley did poorly. We separated the mimic, and we began to care for this new crop, rye, with all the care and love that it desired.

"Will You Walk Upon My Pistols?" Said the Flower to the Fly

paraphrased from "The Spider and the Fly" by Mary Howitt

For an animal, the desire for food is compelling, as is the desire of the female to place her offspring in a good home. Many plants have learned how to use their desires for these resources to their advantage. The century plant (*Agave Americana*) is big—about the same size as the carcass of a large animal—and its putrid smell attracts carrion feeders, especially flies (426). These flies expect to lay their eggs on a piece of rotting flesh, where their babies can feast on the bacteria that grow there. Instead, they find themselves taken in by a very clever plant. As with so many other resource mimics, any eggs that may be laid there will be doomed.

The family *Droseraceae*, sundew and pitcher plants, have more sinister motives. Both of these plants produce heavenly scents for flies and other insects looking for a needed resource. That resource may be sex, food, or a haven with food for their eggs. When the fooled insect gets to the plant, the motives of the plant become clear, but it is too late for the insect. The plant traps the insect, and it then digests the hapless creature to obtain a resource it needs for its survival in its unforgiving environment.

Pitchers and other carnivorous plants usually live in bogs. Bogs are particularly harsh places for plants to live. They are often low in the nutrients needed to photosynthesize and to thus sustain life (427). For plants, one critical nutrient in short supply is nitrogen. Other minerals that can be in short supply include phosphorus and potassium. In plants, meat-eating has arisen independently about nine times (428). About five orders and six hundred species of plants have solved the problem of getting enough nitrogen by digesting insects (429; 430).

Carnivorous plants seem to have little interest in the sugars, amino acids, or other complexes that the bodies of the insects provide. Their primary interest is the nitrogen.[35] Different meat-eating plants use a wide range of methods to trap their prey including pitfall traps and flypaper, cage traps, and vacuums. Each of these plants benefits from tricking the prey, luring it into that trap. As unbelievable as it may seem, the various traps devised by plants strongly resemble the traps developed by human hunters.

The trap of the sundew is simple. The sundew plants produce a sweet sticky substance on their leaves. This promise of a sweet treat attracts small nectar-feeders to these little plants. The insects then become stuck in the gooey mass of sugars and digestive enzymes. Like quality ketchup, this fluid becomes thicker as the insect struggles while the

---

[35]At least some carnivorous plants do use some of the complex compounds in the body of the insects. This information is a giant step away from the old dogma of the plants having only interest in the nitrogen. Nitrogen, however, remains their primary target.

plant begins digesting it alive. The digestive enzymes quickly do their work, sealing the fate of little gnats or other small insects.

Pitcher plants, such as *Cephalotus*, use a stunning mix of different strategies. It has a pitfall with slippery sides. It incorporates ultraviolet patterns into its trap; these UV signals attract a wide range of pollinators (431). Once an insect gets captured, it has trouble crawling out of the liquid that the plant collects at the bottom of its slimy dungeon.

This deadly fluid contains a mix of water, digestive juices, microbes and bait. Along with the promise of nectar, this bait may hold a promise of meat. Into this liquid[36] mix, the pitchers inject repurposed stress hormones. Plants usually release such hormones when a herbivore munches on their leaves. Its scent attracts predatory insects that will attack the pesky herbivore. Although this hormone has become repurposed for digestion (432), the plant may also use the rich scent to attract both herbivores seeking a good patch of food and their carnivores who follow the alluring aroma of foraging food

One finds the same kind of trap in a different family of plants for whom a meat dinner is not the goal. Instead, they catch and trap small insects as a service to another insect, the assassin bug. The predatory assassin bug will eat the trapped insects, and the plant will get its nutrients from the feces of the predator (433).[37] By using the pre-digested waste, the plant can obtain its nitrogen without wasting its

---

[36] Sundews, and some pitchers, such as Nepenthes *rafflesiana*, have digestive juices that thicken around their prey as it struggles (653). This property, shear-thickening viscoelasticity, is yet another mechanism of death employed by these plants.

[37] A wide range of plants have sticky surfaces that can trap small insects, but botanists have only investigated a few of them for carnivory (654). Researchers believe that most of these adhesives serve other functions.

own energy making digestive enzymes.

In another example, the tobacco plant catches troublesome small insects on its leaves. Larger predatory insects will come and eat these plant parasites and leave droppings as they go. The tobacco plant and the predatory insect thus have a relationship similar to the one enjoyed by the fish and cleaner shrimp.

Other carnivorous plants produce snap traps that lure insects to a sticky platform, which they then enclose with their leaves. Venus flytraps (*Dionaea*) and the aquatic Waterwheel (*Aldrovanda*) have these death chambers.[38] Pitcher plants and cobra lilies use a different trap: a pitfall trap. These plants also use a visual or chemical lure to attract their prey, usually flies. Plants may bribe their victim with the promise of nectar. For others, the visual appearance of the plant will draw the dupe into the pitcher. Once inside this deep cavern, the insect drowns in a pool of water. Some pitchers have mutualist bacteria that digest the prey. Others secrete their own enzymes. Still other species may harbour insect larvae that will eat the trapped prey (434). When the plant depends upon one of these mutualists, it uses the waste products produced by them as the source of its nitrogen.

Like animals, plants lie. Fungi lie. Every living thing lies—even things without brains, without a means of verbal communication, and without

---

[38] For years, botanists believed these two deceptive carnivores developed their meat-munching habits independently of each other. The flytrap has anatomical similarities to the sundew, but the waterwheels more closely resemble aquatic bladderworts (Utricularia). Other botanists suspected that these complex traps evolved once, and later diverged to look like sundews or bladderworts because their ancestors invaded different habitats. When botanists looked at the nucleic acid codes of these plants, the results supported the latter view (655). Snap-traps now appear to have evolved only one time, about sixty-five million years ago, toward the end of the age of dinosaurs. The aquatic waterwheel began to resemble other aquatic plants because of other environmental pressures.

a central nervous system. Deceit helps organisms survive by increasing the odds that they will get their own needs met.

Chapter 10

Behavioural Lies: Surviving Youth and Adult Desires

"Fear is a liar, he will take your breath, stop your steps"
~ Fear is a Liar by Zach Williams

Fight or Flight? No, Freeze!

A few years back, I did some fieldwork in South Dakota. My partner and I had focused on a site that we thought would contain a vein of 63-million-year-old amber. After hours of careful searching in oppressive, unrelenting sun and heat, we prepared to leave empty-handed.

As we put the tools away, I turned around and looked behind us. There, not twenty feet away from where we had spent the better part of the morning, I saw a baby mule-deer in a shallow depression. Its spotted brown body and lack of motion gave it such efficient, incredible camouflage that neither of us had even noticed it until then. It continued to remain motionless as we snapped its picture with our cameras. We may not have found any amber, but we certainly left feeling very satisfied with the morning's experience.

On an earlier day, a group of about five of us encountered a small herd of these deer while combing two sides of a canyon. Without any warning, the deer sprang into action from the middle of the canyon and ran ahead of us. Soon, they disappeared around a corner. Until the herd moved, we had no idea that they surrounded us.

The baby mule-deer and the adults illustrate a point: Both behavioural lies and behavioural camouflage can differ between phases of a life cycle. The behavioural camouflage of freezing or inactivity occurs in both adult deer and their young. The young mule-deer in the above example did not move, even though we remained within feet of it for hours. The adults, however, whose stride covers more ground, ran when they believed themselves surrounded by what they must have assumed to be a hunting pack. Both the adult deer and the young deer had no way of understanding our benign intent; they only knew we looked and behaved like a pack of predators. Their responses differed because of their respective ages.

For this difference to happen between phases of a life cycle, chemical changes must occur in the body. In the case of the deer, they develop different sensitivities to the hormones and neural signals that control the "fight-flight-freeze" response.

As a mule-deer moves from infancy to its youth, it experiences one of these sudden behavioural shifts. It no longer hides; it runs. It crosses a behavioural boundary that arises as two such contradictory forces push it in opposite directions. The payoff from the safety gained by running increases as its legs lengthen and its endurance increases. The safety gained by freezing decreases as its size increases and its coat colour changes. These change in the deer result in a sudden, discontinuous change in behaviour, a mathematical "catastrophe."

In this context, the term "catastrophe," does not carry the same meaning as the day-to-day term. It does not imply a horrific event. In mathematics, the term describes a system where changes happen abruptly rather than smoothly. Behaviour suddenly flips from one form to another form. Sometimes, a very tiny tweak in the system triggers the change. The fight-or-flight response is a classic example of a catastrophe. So is a sudden change in stock-market prices, the sudden formation of a tornado, or the rapid explosion of a marine worm to release her eggs.

Imagine a piece of paper with upward pressure applied to both of the two shorter sides. The resulting pressure will cause a cusp. If you keep pushing, the cusp will tip over, and form an overhanging fold. Now, envision yourself at the top of the cusp, or on the edge of the fold. One direction represents one behavioural choice. The other direction is a different behavioural choice.[39] For our human fight-or-flight response, that thin edge between the two choices is the "freeze" zone.

The decision to flee, to fight, or to freeze comes from a region of the brain known as the amygdala. The amygdala controls many of our core survival reactions because it receives information faster than the thinking frontal lobes of the brain. Our amygdalic reactions often

---

[39] Try this experiment: Take a credit card (or other plastic wallet card) and put a slight pressure on it lengthwise by pressing with your fingers on both sides to make the card bulge upward. Now gently apply downward pressure with another finger onto the top of the cusp. Slowly increase the force pushing downward. Rather than a smooth transition, you will discover that the cusp suddenly jumps down to create a cup below the plane between the fingers. If you now repeat the process and push up on the nadir, you will again see a sudden jump upward to form a cusp. Voilà! You have just caused a catastrophic change in the system.

precede what our frontal lobes decide to do. In a situation that the body considers critical to survival, we rely on our amygdalas to make the right move along our internal map.

The sizes of the different parts of this behavioural plane represent the probability that an animal will engage in the corresponding behaviour. Young deer have a much larger freeze-zone on their behavioural plane than we do. The specific combination and proportion of hormones that move us rapidly from flight to fight have a more muted effect (i.e., larger freeze zone) in baby deer than in humans. For that reason, deer freeze, and we run.

In humans, we talk only about a fight-or-flight response because our tree-dwelling ancestors would not have received much of a benefit from freezing. A much stronger benefit would have come from darting up the tree as fast as their legs could take them, or in the absence of such a safe haven, by turning and facing the enemy head-on. Very little of our behavioural experience involves freezing.

For these reasons, fight and flight dominate our response to danger. When we discuss the flight-or-fight response, we often neglect our third option: do nothing. In us, the freeze-response is a narrow boundary zone between our two dominant responses. We get frustrated in a crisis when our political leaders appear to be doing nothing.

On November 4, 1979, a small group of militants took over the United States embassy in Tehran, Iran. They held 52 embassy-personnel as hostages in an incident that would last until Jan 20, 1981. When U.S. President Carter did not immediately send troops to obtain the release of hostages held there, he lost face with the American people who wanted "something" done. The pressure resulted in an ill-conceived rescue attempt that led to several deaths among the rescue workers.

The crisis contributed to his loss to Ronald Regan in the next election. When President George Bush (Sr.) did not immediately respond to the Exxon-Valdez disaster, people quickly criticized him for his inaction, just as they criticized President Obama for his initial inaction in 2010 when oil plumes contaminated the waters of the Gulf of Mexico. George Bush (Jr.) received similar criticism for his delayed response to Hurricane Katrina.

In contrast, the federal government of Canada invoked the War Measures Act and deployed the military in response to the kidnappings of two government officials in 1970. On average, the Canadian people initially supported the swift action. Only later did they target the government with criticism for a hasty response. In the United States, people felt frustrated at the survival of Osama bin Laden for nearly ten years after the 11 September 2001 attacks on New York and Washington D.C. When bin Laden died on 1 May 2011, many Americans had an initial sense of relief and a feeling that it was justice too long delayed. They liked seeing the action taken. The following day they already began to focus on the potential downstream consequences of the action.

Today, many Westerners praise the passive resistance of Mohandas Karamchand Gandhi as a critical element in the Indian fight for independence, but they forget that his movement fell on a backdrop of action by people such as Barin and Aurobindo Gosh, the Jugantar party, and the Anushilan Samiti. In Ireland, people still remember the actions of the Easter Uprising of 1916, which resulted in the deaths of over 400 people, as a turning point toward independence. As a species, we dislike inactivity, and we demand a rapid, active response by our leaders. We may complain later about those swift responses, but we do like action in the here and now.

Deer, on the other hand, have a much larger region of their plane of possible behavioural responses dedicated to the freeze option. In young deer, this option is more pronounced than in adult deer. To tip the balance in favour of freezing there must be internal regulatory changes in hormones or neurotransmitters. To induce a freeze response, the brain of the fawn must suppress those command centres in the brain that move the adult toward action. To induce a flight response, the adults must become sensitive to the pathways that fawns ignore and insensitive to the commands that make fawns freeze.

As they age, the bodies of many species change their physiological responses to excitatory compounds, such as adrenalin (an excitatory hormone) and dopamine (an excitatory neurotransmitter). They can change their titration with these messengers, or they change their sensitivity to these compounds. Alternatively, they may become less sensitive to inhibitory hormones and transmitters.

Like any other trait, nature selects for the behavioural camouflage of freezing when it leads to enhanced survival. Young insects, like young deer, can benefit from behavioural camouflage that they can shed as they grow older. When insects moult between larval stages or from their larva-hood to adulthood, they are extraordinarily vulnerable. They lose their old exoskeleton as they emerge from it. Their new outer skeleton, which will eventually afford protection from many predators, begins as a soft outer coat rather than a hard one. It can take from minutes to hours for their new armour-plated exoskeleton to harden.

During this eclosion time, the wings cannot function properly, and their muscles have no hard surface to give them leverage. The helpless insect can neither fly nor walk. Like the young deer, whose little legs could not outrun a predator, most insects opt for a quiet, nearly motionless existence during this vulnerable stage of their life cycle.

Very young mosquitoes, known as wigglers, also have to lie through behavioural camouflage to survive. Sometimes, motion can provide a camouflage that is superior to motionlessness. The larvae combine low activity levels with a few other tricks to stay invisible on the water surfaces where they live. Their silvery colouration and wiggling behaviour help the wigglers blend into the shifting silvery patterns visible on top of the water. The combination of all of these bluffs helps them avoid the jaws of hungry fish.

To Bluff or Not to Bluff: A Balancing Act

Not every behaviour of a juvenile or of a defending parent is a bluff. Tricks, such as mimicry, quiescent behaviour, and camouflage lose some of their effectiveness if they are overused or if a target learns to expect them. If there are too many bluffers in a population, the bluff can lose its effectiveness. For a lie to work, the dupe must believe the lie. It can also help to have honest tricks up one's sleeves.

Occasionally, bluffs can turn into a real defence when others join in the same behaviour. Herds of adult bison, cattle, and musk oxen will surround their calves when they are under attack. For an individual parent, the defence posture would still be real, but not necessarily very functional. The bluff would work only if the predator believed that the individual parent could harm him or her. This pathetic defence becomes an impressive one once the individuals join forces.

Similarly, when youths band together, their defences can also become stronger. Beyond collective camouflage, young animals can collectively bluff their way out of danger. Individual owlflies are relatively helpless little creatures. Their mothers compensate for this helpless state by laying eggs in clusters. These broods increase the probability that a larva will find a few of their brothers and sisters soon after emerging. When threatened, the larvae form a tight military formation on their home twig, just like the musk ox do around their young. This phalanx faces outward, with their many jaws snapping in all directions. This 360-degree defence makes the young owlflies a less appetizing prospect to an enemy. Thus deterred, many predators will look elsewhere for less costly prey.

Social behaviours, such as the defensive behaviour seen in the owlfly and the bison, can multiply the impact of a behaviour on a predator. When placed into a social context, a behaviour that functioned as a lie or bluff in one individual owlfly or bison becomes a real force of nature when they work together.

The aggressive behaviours that may be seen as self-deceptive overconfidence when practiced by one individual, can cease to be a lie as the sheer number of participating individuals magnifies the impact of the behaviour (435; 436). Such self-deception may, at times, serve as the foundation upon which an organism can build a strong group defence.

This phenomenon worked to the advantage of the United States after the bombing of the World Trade Center and Pentagon. After the attack, the United States experience a dangerous combination of unity, outrage, and a high level of self-confidence in its ability to mount a retaliatory attack. The eagle sharpened its talons and quickly soared toward revenge. Its overconfidence in the efficacy of its reprisal led many Americans to believe that the war would be over before the start of that summer. Instead, it became the most protracted conflict in American history.

A self-delusion of assurance, however, made the American attack less predictable to their enemies, and it helped to establish a reputation for violent retaliation. Such a reputation can be beneficial in a world that does have potential predators. Humans also use such strategies of confidence contagion when they rise against a superior power or wage war, when they create an impasse during labour negotiations (437), or when they join protest rally or a stock market bubble. Individuals use confidence contagion when they join a support group to change a behaviour. Such delusional group-think can help us change into what we cannot deceive ourselves into becoming on our own.

A predisposition toward accepting misbelief also helps a social organism thrive by predisposing it to work with its group toward survival (438; 439). Self-deception and the over-inflated ego and confidence that goes with it can help us sell the lie to others, (440; 441) and then to ourselves. In some cases, the lie becomes the truth as it begins to work. In these ways, self-deception may function like other easily triggered behaviours that have a similar effect.

The Rigged Lottery

Similar group defences occur in many ground-nesting species. Behaviours that may be a bluff when alone can work for a group. The effectiveness of the combined defence is partially determined by how easy it is to recruit the neighbours into the defensive posture. Prairie dogs typically go underground when a colony member raises a warning signal. When a snake approaches a colony of prairie dogs, it threatens the burrows themselves. Unlike other predators, snakes can slither into these underground homes.

When a prairie dog sites a snake, he or she gives a different signal, and a behavioural catastrophe occurs. The alternative alarm triggers a different fixed pattern of behaviour. The little rodents will charge the snake and kick sand at its face. As a group, they protect their burrow with behaviours that would be suicide for any individual attacker. The benefits that individuals derive from the combined defence outweigh the risks assumed by any one prairie dog (442).

A large number of attackers decreases the chance that the snake will attack any one individual. Effectively, they form a union. Ethologists call this reduced risk the "dilution" effect. By cooperating, the prairie dogs reduce their chances of predation by spreading that risk to a larger group, just as union members, who put pressure on management as a group, reduce their risk for retaliation against individuals.

For the dilution strategy to work, however, the defensive behaviour must be an easily triggered fixed pattern of action in each of the rodents. Each rodent needs to behave as if it has an inflated sense of its chance of survival or be unaware of the risks. As an individual attacker, a dog would be dead. As a group member, it is likely to survive. Even if the predator makes the unwise decision to attack the colony, the group reduces its chance of success and dilutes the risk to each member of the colony.

Ducks and geese also use the dilution effect to protect their broods. Although most species of waterfowl lay only a few eggs, it is common to see these parents with twelve or more ducklings behind them. Adults willingly accept orphaned babies into their families. Ducklings will also abandon a small clutch of brothers and sisters to join a larger clutch. Because these babies can feed themselves, the adoptive parents have little extra work in rearing a few extra chicks. Their natal babies benefit from their adopted brothers and sisters because an attacking predator now has a chance of eating one of the adoptees. In the larger clutch, the adoptee similarly benefits from the dilution effect. Thus, adopting the extra offspring may not be a case of generosity and altruism on the part of the adult duck. Instead, it may be a willingness to use the adoptees as substitute fish food.

Who's Your Daddy?

Among animals, females are usually confident that an egg or offspring came from their bodies. Males have no such biological guarantees. As mentioned elsewhere in this book, this difference contributes to distinct differences in parental behaviours among males and females. The more uncertainty of paternity within a species, the less likely the male will spend time helping to raise the offspring. Females can increase the odds of male involvement by making it harder for males to mate, thereby increasing the cost of mating.

Many species also experience rewards for fostering someone else's offspring. Fostering within a species occurs in over one hundred different mammals and birds (443). If the benefits of accepting the changeling exceed the cost, then adoptive behaviour is favoured by natural selection. Surrogate parents may reap benefits because the fostered animals may carry some of the genes of the adoptive parents, it may help keep their offspring safe, or it may help gain access to the animal equivalent of a babysitting circle. In king penguins, for example, adults regularly feed offspring other than their own (444). Passing excess food to other babies imposes little cost and increases the survival rate of its related flock-mates. If most of the feeders had no offspring of their own that season, their only chance for any genetic success lay in helping their relatives in the flock.

Parents are not alone when it comes to manipulating the parent-child relationship. Offspring also make choices. In penguins, babies often switch parents. Penguin parents take turns nurturing their offspring and keeping them safe on a "nest" that is actually the top of their feet. If a parent is unable to provide a good standard of living for a chick, the chick will dislodge itself from its parent and run to the feet of a more promising neighbour. There, they behave as if nothing unusual has happened. The new penguin parent, who is not a direct genetic relative, will usually adopt the newcomer as its own.

At first glance, there does not appear to be any real advantage in the adoption for the adult penguin. The chick does not carry the genes of the new parent. At best, this new chick is a distant relative by virtue of being in the same flock or the same species. Rearing that baby seems to be a waste of good reproductive time. So why do the adoptive parents treat the intruder as one of their own?

A partial answer to the adoption paradox seems to lie in our old friend, the concept of the fixed action pattern (FAP). Recall that FAPs channel actions to a pre-set path from which it is very tough to veer away. For the penguin parent, the presence of a baby triggers parenting behaviour. The FAP is so easily set in motion that even a baby from a different species can sometimes trigger parenting behaviour (445).

In the harsh conditions in which the penguins live, every egg, every baby, every chance to pass on the genes becomes enormously precious. Genetic survival depends upon easily triggered, fixed parenting. The typical cues to elicit parenting now fire from hair triggers. For the penguin, the cost of making a mistake and expelling a biological offspring from care outweighs the cost of taking on an extra mouth to feed. The successful deception of the scamming baby is possible because of the intensity of the parenting FAPs and their shared goal of genetic survival. The trickster baby penguin has broken the parenting code—the same kind of code that triggers the parenting FAP in cowbirds and cuckoos. The only significant difference is that the parasitic cowbirds and cuckoos (as seen in Chapter 4) broke the code between species rather than within their species.

Like the penguins, humans also have small clutches of children, usually giving birth to only one offspring at a time. This similarity in reproduction correlates with similarities in behaviour. We too have strong fixed action patterns toward our children, and we too adopt. Most human societies adopt within extended family groups.

In traditional societies, fostering within a family was common and strict rules governed which relative could foster which child. Usually, a sibling, a grandparent or another relative adopted the child whose parents could not properly care for it. For example, among the Hopi of the American Southwest, adoption of a child would traditionally fall to the mother's line. If no direct family existed, adoption occurred within clans.

Within-family fostering can work to the genetic benefit of the adopter. On average, grandparents will have 25% of their genes in common with their grandchildren. Should a biological parent be unable to rear a child, the grandparents can prevent the loss of these gene-carriers by rearing the child themselves. Siblings and cousins will also have investments in this adoptive child.

Even if the parent can raise the child successfully, in-family adoptions and familial support of child rearing may help a relative to dedicate more energy toward having more children. In this manner, the adopter increases his or her indirect contribution of genes to the next generation through the efforts of a relative. In this way, natural selection works through the mutual support of kin. (For more information on human adoption, see Appendix B: Adoption Behaviour).

Humans (and many other primates) have a very long period of dependency. Consequently, for females, each child represents a greater part of her total genetic potential than a child represents for a male. Nature has heavily weighted our fixed parental behaviours toward avoiding the loss of a natal child by mistake.

As adults, we respond to baby characteristics. We love these traits. We think of big heads and eyes as "cute." Toys and cutesy-pie statues that have these physical traits sell well in stores.[40] Sharp piercing sounds, like the cries of infants, make us uncomfortable and trigger a desire to stop the screaming by rendering aid. In these manners, the look, smell, and sound of babies tend to trigger these FAPs in most adults. Babies—and things that look like babies—tap into our innate coded behaviour patterns; they instinctively know how to take advantage of these common adult attributes.

---

[40] Manufacturers market these cute-faced toys primarily toward women. Women tend to respond more deeply to cute, baby-like features than do men. Although many women want to collect small statues and toys with big eyes and heads, men more commonly harbor a secrete desire to break them.

Human FAP-responses to infants are strong enough to allow other organisms to crack the code. We respond to large eyes and heads of animals the same way we respond to these cute features in human infants. In some cases, we selected those features in our domestic animals (446; 447). Among lap dogs, small poodles are particularly good code-mimics. A miniature poodle is about the size of an infant. Its long ears have the same general shape as long hair on humans. The high-pitched voices of the miniature and toy poodles also mimic the pitch of a human infant.

The adorable animal benefits from our impulses toward nurturing and affection. Its very image gives us a small dose of a hormone, oxytocin which fosters warm and fuzzy feelings. Cuteness, in essence, manipulates us. It releases a fixed behaviour pattern that makes us look upon those who are cute with love and care (448; 449; 450), and a tendency toward leniency for infant behaviour (451; 452).

Our cute-response to these animals elicits more careful behaviour as we interact with them, similar to what we show when handling an infant (449). For these reasons, with a few exceptions, women tend to like miniature poodles and other small lap dogs.[41] For this same reason, men often take a strong dislike to the small breeds of dogs. For them, the mimicry may tap into the FAP resentment toward someone else's baby. Small dogs benefit from their resemblance to human babies more than they are harmed by it. Like the baby penguins, little dogs are happy to

[41] Culture also places limits on what objects we can see as cute. For example, some cultures do not consider dogs as clean, socially acceptable animals; people in such cultures would be less likely to perceive small dogs and puppies as exceptionally cute. The phenomenon is not limited to cute dogs. In general, women like "cute" things more than do males.

take advantage of parenting FAPS with whatever species might oblige them.

In many societies, the same innate familial trigger that tells us to adopt a cute puppy or another person's child also tells us to feed the child who comes into our home alongside our child. For most of human history, the children with whom our children have associated—the children whom we were likely to see—would have been kin. We may consciously recognize the child's friend as an outsider, but our social and biological histories still tell us that this child may be kin or may be of some other benefit to our little bundle of genes.

In all these cases, the behaviour and patterns are not intentional deceptions. Systems already in place trigger the behaviours to prevent the loss of reproductive potential. These behavioural responses are internal lies that we are more than willing to accept.

Young Women and the Centrefold Effect: What Men and Women Need

"Power is the ultimate aphrodisiac."
~ Henry Kissinger in *The Sayings of Secretary Henry Kissinger* compiled by DuPre Jones

Across cultures, heterosexual men show a desire for the same kind of sexual object: fertile-looking females with highly symmetrical faces (453; 454). This feature may signal health and a life less riddled with disease than average. Body proportions—the desirable hourglass figure with the waist narrower than the breasts or hips—also signal fertility. The desired measurements may differ around the world, but the preference for the relative proportions remains about the same (455;

456). Men also favour a youthful, or neotenous, appearance, which is another marker for female fertility (457; 458). Across cultures, clear skin and evidence of health usually rank as attractive. If one checks the centrefolds in typical men's magazines, these features dominate.

Human females have a dramatically different set of reproductive needs. They strive to get the best sperm for their eggs, but they also strive to provide the best opportunity for their offspring. From a Darwinian perspective, one would expect human females, on average, to be more concerned with material security than males because material security leads to a better chance for the survival of their high-investment offspring. From an evolutionary perspective, the male attraction to material goods occurs because the accumulation of status improves their reproductive potential: it attracts females.

This attraction to physical resources cannot exist on its own. Human females must also look for markers of fertility and genetic resources among males with money and material goods. No matter how successful the male, he must also show signs of fertility. In males, fertility entails, at the very least, a high-enough level of testosterone to make sperm.

Human females might have an interesting biological solution to this problem of needing men who are both fertile and financially stable. Some—but not all—studies suggest that human females find different features attractive at different times of their fertility cycle.[42] At peak fertility, women seem to prefer facial features that indicate high

---

[42] A meta-analysis by Jones. Hahn, and Fisher (657) failed to find evidence to support the idea that the preferences for more or less masculine features varied across a woman's hormonal cycle. Although this study stands in contrast with much of the literature, its lead author, Benedict C. Jones, previously reported the presence of the phenomenon. This about-face should be taken as a credible counterexample. The emotional sensitivity of the material from multiple authors could also contribute to the problems of replicability.

testosterone levels and therefore fertility; at other times, women prefer less angular features that portray a less aggressive, less testosterone-laden male (459; 460). They show a similar fertility-biased preference for the odour of dominant males during their fertile phase, but they prefer less dominant "smells" during their infertile times (461). These odour preferences can also vary with relationship status (462). Fertile women in stable relationships have stronger odour preferences than fertile single women, possibly because those women in stable relationships are much closer to having offspring than single women. Thus their *immediate* investment is greater. When progesterone (the hormone that maintains a pregnancy) levels are high, the commitment to a relationship also may also increase (463; 464).

Women also show a preference for the scent of men with immune-system profiles that have moderate differences from their own (465; 466; 467) possibly because such males are less likely to be close relatives. Also, offspring conceived with such an immunological complement would have a more varied immune response and therefore a better chance of survival.[43]

Not too surprisingly, men do not seem to share the aversion; they find women attractive despite their immunological background (468; 469).

These data give insight into why men and women may deceive each other in matters of sex. Males and females may be tempted to throw judgement and caution to the wind for different reasons. The lure of multiple women may tempt men because they fight primeval

---

[43] We may have developed a dislike for radically different immuno-profiles early in the history of our species, as we differentiated from our cousins the bonobos. The preference may have steered us toward the more genetically compatible mates in the diverging groups while still steering us away from our immediate kin.

programming to leave behind many and varied offspring. Women may look to men other than their spouse for sex because such men may provide superior genes for their limited number of offspring. By having offspring by more than one male, a woman can deposit her genes into more varied combinations, increasing the odds that at least a few of her children will survive.

Because of these different imperatives, women must solve a complex problem. Their primeval programming tells them to gain resources from a successful male. It also tells them to access quality and varied genetic material that he may be unable to provide.

Those females who do engage in such extracurricular activities to find better genes must also protect their offspring from repercussions should a cuckolded spouse discover the deceit. Many male mammals, including humans, respond poorly to offspring that are not likely to be their own. When a new male lion or gorilla, for example, assumes leadership of a troop or pride, he typically will kill the still-suckling babies of the previous male. Human males will frequently favour their children in a blended marriage to those offspring brought into the family by a spouse.

Women must account for these more violent aspects in the evolutionary history of human male behaviour. If a female can understand or predict the behaviour of a potential mate, she can make better selections that may aid her future children.

Compared to other primates, human males have an unusual attachment to their offspring. From the perspective of traditional evolutionary theory, this attachment does not make any sense. One can argue that men should not form attachments to their offspring because through most of our evolutionary history, males would not have understood that a child was the result of copulation with a particular mate.

If we once behaved like our promiscuous bonobo cousins, our closest biological relatives, then our distant ancestors would have had no way of knowing whose baby was whose. Without a single dominant male accounting for the preponderance of reproduction in a troop or smaller social unit, there should be little selection pressure for a dominant male in a social unit, like a family, to favour the survival of specific offspring.

Humans, bonobos, and chimps, however, do share one other trait that begins to explain why our males may tolerate offspring. In all three species, the females join troops of males. Males tend to remain within the community of their birth. Males in these troops have family ties to each other. In chimps, the dominant male, who is often the father, is at least a cousin to the other males. They have a genetic stake in his offspring, albeit a small one. The male bonobos probably have had sex with most of the females in their troop. They have the potential of being the father. Also, female bonobos might join forces to kick his butt if a male attacks one of their offspring. In the Pleistocene, the era in which many of our human behaviours took shape, humans also lived in similar small bands of brothers, each of whom benefited from the survival of their young.

Why Do Humans Have Devoted and Affectionate Fathers?

"Bassanio: So may the outward shows be least themselves;
The world is still deceived with ornament."
~ *The Merchant of Venice* by Shakespeare

Females may have actively, if unconsciously, selected males that showed traits that made them good fathers (470). They may have chosen males who were more likely to be kind to the offspring and who would sacrifice for their children.

On the surface, such decision-making may seem unfeasible. One might assume that an ability to predict future behaviour might be all but impossible. The solution to this problem might reside in a combination of two different factors. First, the primordial women did not need to be right 100% of the time. They just needed a mechanism for prediction that would, on average, help them to choose mates who had the desirable traits. This subtle bias toward kinder and gentler males would, over time, help domesticate male behaviour. The other factor would lie in the innate but clever tricks used by human females when they choose these desirable mates.

Inter-sexual selection is a type of evolutionary pressure in which one gender differentially mates with another gender based upon preferred characteristics. Female damselflies may prefer males who can maintain a territory with the resources they need for their eggs. Consequently, they select for territoriality in the damselfly males. Peahens chose males with large tails, and prairie chickens choose the best dancers. Males of these species put their best foot or wing forward, trying not only to look their best but also to block or to sabotage the efforts of their rivals.

Women may make similar kinds of choices, and they must also see through the often misleading signals and competitive tricks of the men. Unlike the peahens, who look for the best variation in tail feathers, human females focus much of their energy on behavioural variation.

In humans, females actively choose males with desirable behaviours. Among the tools that they use to accomplish this task are the differences in appearance between men and women. Women use this slight dimorphism to help them choose mates who will be good to their offspring. Compared to the other apes, humans retain far more childlike features into adulthood.

These juvenile or paedomorphic features of women are replete. They include a flattened face, reduced body hair, and a characteristic eye-fold. The position of the head relative to the spinal cord, positioning of the orbits of the eyes, and even dental features all differ from those of men (470).

Female humans keep many more childlike physical traits into adulthood than do men (458). Across cultures, men find these paedomorphic features attractive (458). Women remain smaller than men, and physically they are slightly weaker. They tend to have less facial and body hair than the males, and their voices retain a higher pitch. In short, human females experience fewer changes in their appearance as they move from childhood to adulthood.[44]

Women watch. They see how males treat them and how they treat other females and children. In those human cultures that permit female choice to run its natural course, females will include these factors as they weigh their decisions about potential mates (471). The youthful body shape of females sends subtle cues to which males respond. Because males, in their turn, have come to prefer neotenous females, they also tend to treat youths well.

---

[44] As a species, both men and women retain many features of the juvenile into adulthood, far more than the average primate. We keep a flatter face and a higher brain-body ratio than do the apes. We stay naked, a trait that most mammals lose soon after birth. The neotenous traits of which we are most proud is that we continue to take joy in learning and in playing throughout our lives. Although both men and women are highly social, close male friendships are rare in other adult primates who must compete for females. Human males not only make male friends well into adulthood, but most of them remain very homosocial, with a strong desire for male-male social bonding.

Females take this tendency one step further. After assessing how the male treats women and children, they select men who are more tender to both. They use the response of the former to determine their behaviour to the latter. By choosing mates based on their needs and the needs of the offspring, female humans have heavily domesticated males. Today's human male results from this intermittent but powerful selection pattern used by females for thousands of years.

The centrefold effect is one way to explain the unusually good behaviour of human males toward children. Even if the centrefold effect—the deceptively desirable paedomorphic appearance of the female—is not solely responsible for the domestication of man, the typical good behaviour of the human male toward children suggests that for at least a fraction of human history, our males have experienced selection pressure to be good fathers.

Sometimes the system runs amok. Males may use their collective strength to reduce the choice of females. Sometimes the attraction to childlike traits in the females may extend to an attraction to children themselves. Sometimes the cultural and familial experiences teach a woman that she needs to use her set of paedomorphic traits in a manner that is maladaptive or emotionally infantilizing. Most of the time, however, the system works amazingly well.

Chapter 11

Drugs and Chemical Lies

"Julie sweated as she walked the stairs to the hotel room. She had had a hard day, but everything would be fine now. Joe had an extra set of needles and works, and he readily shared them with her. She had managed to score some heroin down by Broadway and Wilson. She was not in the room for five minutes before she had her arm tied off and the drug cooking in her works. Soon, she sucked the H into the needle and shot it into her vein. Within moments, the pain went away, and a wash of blessed relief came over her. Once again, the world was set right."

~ "Julie's World" by Circa24, unpublished

This chapter is dedicated to John Cebuhar, 1950-2018,
my friend and colleague.

Most people would call Julie an addict and, with that statement, dismiss both her and what her experience has to teach us. Our brains and our culture help us build an image of Julie as being in a dingy flophouse motel, surrounded by roaches and dangerous characters. We tend to picture Joe as an unsavoury character, the kind of guy who runs "shooting-galleries" where addicts congregate to inject themselves into a drug-induced stupor. We see them both as wasting their lives.

That image is not the complete picture. Others are out there. Many opiate addicts never set foot in a flophouse, never lose their jobs, and never run through their money. Julie and Joe may very well be the flophouse duo. They might also be Wall-Street professionals who meet at a posh hotel for their after-work pick-me-up before heading home. They may be Beverly Hills socialites who meet in an alley behind the hottest club in town or a well-to-do couple who hook up at the Fairmont in Whistler, B.C.

We do not like the latter picture of the functional addict (472). Despite the accompanying prejudicial image of junkies, the ingrained picture of the strung-out street-junkie has far more value to society. It gives us an image of what we might become should we ever start using heroin. We have corresponding images of people who abuse alcohol, cocaine, marijuana, and other drugs.

The dissonance created in our head by the juxtaposition of an image of a drug addict and the image of a successful human being makes us squirm with discomfort, so we tell ourselves that functional addicts are rare exceptions. We prefer to punish the drug users rather than to understand the physiological, social, and psychological patterns that could lead to more effective treatment (Flavin & Paltro, 2009).

"Sometimes, You Get What You Need"

~ "You Can't Always Get What You Want"
by Mick Jagger and Keith Richards

Do not misinterpret me. I believe that drug abuse *is* bad for you. That is why we call it "abuse." Some drugs, such as tobacco (473; 474) and opiates (475) can cause long-term changes in brain function after very limited use. Prenatal exposure to only a few cigarettes can impair the ability to form plans and regulate one's behaviour well into adolescence and adulthood (476). Others may take much longer to inflict damage to the body.

Unlike arsenic and cyanide, psychoactive drugs do not do their damage quickly, and not all drugs do universal damage to their consumers. Most people can use alcohol in moderation with few consequences. Thanks to their genetics, many people have a physiology that predisposes moderation in alcohol. People with a slightly different tweak of their genetics or physiology will binge to the point of self-harm (477; 478; 479; 480; 481). Some people use marijuana as a recreational drug, and others at the opposite extreme have their lives destroyed by it. These differences may also have a genetic basis (482).

Variations in the genetic code control much of our fatal attraction or our revulsion to psychoactive drugs (483). These differences in our base level of attraction interact with the changes wrought by the substances once they are in our bodies and interacting with our varied physiologies (481). For most people, there is no way to know where they fall on the spectrum until after the fact. I personally subscribe to the philosophy that starting to drink or use drugs is a bad way to find out which of these body types one has.

For years, we have lied to teenagers about drugs. We have portrayed all psychoactive drugs as though they were cyanide or arsenic. We have worked with the social lie that claims *everyone* who uses drugs will "hit bottom" and destroy their lives with drug use. However, teenagers are lie-detection machines. They stand in a position of being biological near-adults with no social power. Like other disenfranchised groups, they depend upon their powers of observation to manoeuvre through a world where the dominant primates (the adults) sometimes lie to them supposedly "for their own good." However, children and teens have brains, and they use them.

The result of our well-intentioned portrayals has been dismal. Our society has reared several generations of teens and young adults who know that the propaganda does not match the immediate day-to-day reality of drugs which they see in their friends (484; 485; 486; 487; 488). They sit side-by-side in the classroom with the kids who smoke pot or who drink on the weekends. Sure, they also see the kid who comes to school stoned out of his mind, but teens are perceptive enough to know that people have a range of reactions to drugs.

Parents, television, law enforcement, and drug-prevention programs provide a mishmash of true and false information. It becomes next to impossible for someone to sort through the mix of good and bad data. The average person takes the information she receives; then she sorts it based on the perceived credibility of the source. She will then reach a decision that achieves a degree of consistency between her beliefs and behaviours.

Psychologists know this desire to unite our thought processes into a cohesive picture as a desire for cognitive eusonance. Movement away from that eusonance is known as cognitive dissonance (489). The desire for eusonance happens across age groups. We find excuses for disbelieving many an unwelcome message if doing so is easier than changing our behaviour or upsetting our peer group.

While trying to achieve this eusonance, our brains prefer the path of least resistance. When listening to a radio preacher, we must decide if the Reverend Cuthbert really does have a pipeline to God or if he is a charlatan. If we believe in him, we do so not because we have divine inspiration, but because we like his message. We find it in accord with our patterns of thinking, and we believe that God will like it too. Alternatively, we may decide he is a fraud and that God should sue him for libel. We come to that conclusion because we do not like his message; it does not jive with the facts as we see them, and we think God would also be offended.

We usually take the most straightforward route to a decision. We take the path that seems to require the least amount of change in our way of thinking, of doing things. On average, we put forth little effort to piece together bits of information that we may dislike. When presented with a contradictory piece of information, we will try to reduce the dissonance by taking whatever path causes the least upset to our world.

The teenage mind is not that different from the adult mind in this respect. Teens make the same kinds of judgements as they evaluate the truth or the accuracy of a drug message. They assess its impact on their lives and its eusonance with their own experiences (490). They make a decision either to accept or to reject information based upon the credibility of the source. They determine if the source is lying.

Disenfranchised groups, be they slaves, teens, or conquered peoples, often perceive information received from peers as more trustworthy than information obtained from sources that hold power over them. No matter how accurate a piece of information may be, when received from the wrong source, it sounds like party-line propaganda.

Peers and peer experiences can carry more weight with teens than the opinions of adults (491; 492; 493). People who believe they have little power or value (such as pre-teens, slaves, and low-ranking social groups) may accept the word of a dominant group as authoritative. However, as these same people develop a more empowered self-image, they often see those same authorities as manipulative and out of touch with the world in which the less powerful person lives (494; 495; 496; 497).[45]

Teens experience such an emerging identity. If the adult stories about drugs do not ring true, if they do not jive with daily experience, then the teenage mind will perceive all adult information about this topic as potential lies. Adult attempts to "enlighten and inform" teens may be perceived as patronizing or as scare tactics.

In these circumstances, the form of the message will override any legitimate function unless the recipient of the message trusts the sender of the message (486; 498). In essence, our attempts to paint the darkest, most negative image of drugs can have a paradoxical effect. Rather than being made afraid of drug use, at-risk populations may reject the bleak message presented. They will favour an alternative message that better reflects their views and experiences. Among their peers, they see a different truth: drug use does not harm everyone. In short, if the image of drugs presented in drug education does not jive with the experience of the student, then the program will be counterproductive. Efforts to communicate using the worst-case scenario message may merely reinforce the image of the adults as "them."

Part of the communication gap lies in the action of the drug. The chemical is a better liar than the drug prevention "experts." When we portray a drug as Lucifer incarnate, it quickly becomes apparent that we have not given the entire truth. When the message of the drug educator dismisses the less drastic experiences of the student, then the student dismisses the anti-drug message as a con-job. Teens catch us in the

---

[45] Think, for instance, about how those of us who labour for a living view administrators and politicians.

falsehood because we do not also tell them about the side of the drugs that the teens first see: the beautiful lie of the drug.

"It Felt Good to Get Out of the Rain"

~ "A Horse with No Name" by Dewey Bunnell

John asked the assembled mass of young faces "Why do people do take drugs?"
"Because they are stupid," said one young girl.
"Peer pressure," said a boy, parroting the answer he received in school.
"No," said John. "People take drugs for two reasons: to cure sickness and to stop pain."

~ drug prevention seminar, Illinois, ca 1998

If we were to tell kids the truth about psychoactive drugs, we would need to start with the frightening truth that many of the at-risk teens already know: drugs can make you feel good. No one takes drugs to feel bad. People take drugs to cure sickness and to reduce their pain. They take drugs to feel better. This positive effect may take the form of a "high," a euphoria that lifts them from a drab day-to-day world.

For many years, John, a close friend of mine, gave drug-education and HIV-prevention lectures to at-risk teens, to young adults, to prisoners, to heroin addicts, and to the families of prisoners and injection-drug users. He often stepped out of the narrow role that defined the responses expected by his audiences.

John reasoned that humans search for honest communication in a sea of surrounding lies. When something felt true, his audience would listen. In this respect, the at-risk population acted like every other person and every other species on the planet.

For John, lying about drugs was just not an option. This dedication to the truth had nothing to do with any idealism about human nature. He honed his skills while talking with the people who knew drugs best: the people who used them and suffered through their consequences. If he were to tell an untruth to that crowd, he would lose them. If they were to peg him as a liar on this point, or feel that they knew more about drugs than he did, then they would also reject any other message of his as a lie.

When adults talk with teens about drugs, they too often have an audience that is drug savvy. Today, even the teens who do not use drugs, who reject drugs and would never normally consider drugs, may know someone who does do drugs. Often, students do not see the harm of the drugs, and probably consider the anti-drug message given at school to be disproportional to the reality of the substance abuse.

The personal experience of most teens may not include Julie, the street junkie. More likely, it includes Julie, their study partner or the cool kid in drama who also happens to do pot or ecstasy on weekends. When we portray only the end product of substance abuse—the destitute, homeless, and mentally ill person—we do not feed into the experiences of the world in many schools. We lose the audience for whom the late-stage issues of substance abuse are not part of their experiences.

This difference in experience may also explain why some drug education programs, such as D.A.R.E., are successful in some settings but not in others (499). D.A.R.E.'s stark portrayal of substance abuse and gang violence may ring true if the teens have models for these experiences in their lives. The same presentation will sound like overkill if presented to students whose only personal experiences with drugs is the kid two desks over who sells his Ritalin to the kid three desks over.

When we tell half-truths in an attempt to maintain social norms and values, we risk losing both our integrity and the maintenance of those norms. We have to come clean with our audience if we have any hope of getting their attention and of keeping their belief in us.

For drug education, that means it should benefit the goal of the educator if he or she validates the experience that teens have with drugs. We must acknowledge the physiological impacts of psychoactive drugs. These results include the short-term benefits that people perceive as well as the long-term consequences that remain more hidden.

We need to concede that the world of drugs observed by children, teens and young adults is a genuine part of their social existence. When a person takes drugs to alter his brain function, he chooses a drug that will make him feel how he wants to feel. Psychoactive drugs can alleviate mental suffering and pain. They can take away stress, make us feel good, and provide us with a sense of power and love. They can help us sleep, help us stay alert, and provide us with exciting and unique experiences that can alter the way we perceive the world around us. Psychoactive substances can seem fun, stimulating, and rewarding. After all, as John said, that is why many people take drugs in the first place: to feel good.

To deny the pleasurable aspects of drugs is to deny the evidence provided by the eyes of the observer. You are telling the person, "Look at me. I know about drugs and everything in your experience is fraudulent." Their mammalian lie-detecting systems will take in what you have to say, process it, and spit it out with the rest of the mental trash.

A know-it-all attitude and presentation annihilates credibility. It immediately creates dissonance between individual perception and the words of the so-called expert. No animal will accept a message that goes against its perception of the world. That is why animals spend so much time perfecting their lying skills and their lie-detection skills.

In the 1970's song "A Horse with No Name," Dewey Bunnell created much controversy between generations. He and the other authors of the song have long denied its inspiration by drugs. Still, the lyrics take the listener on what can be perceived as a prolonged heroin experience. The key, the mood, and the lyrics blend to give a sense of empowerment and awe, followed by fear and loathing. When the person finally realizes that the desert is dangerous, it has already consumed him. All that remains for him to do is to lament the beautiful disguise of its original form.

As mentioned earlier, two points hinder our attempts to explain the dangers of drugs. As a society, we seem to have an inability to tell the whole truth when it comes to psychoactive drugs. Second, the drugs themselves tell their own story to the user and his or her friends. The deceptions told by the drugs—the pleasurable and satisfying effects of the drugs—also make the dangers hard to perceive. It is the pleasant deceit that makes people take the psychoactive drug, not a desire to become enslaved by a foreign substance.

Psychoactive drugs work on our bodies because they mimic substances already present in our bodies. They convince the brain that they belong there because the brain sees them as chemicals that it already produces.

Some psychoactives look like pleasure chemicals; others look like chemicals that make the body sleepy, wakeful, calm or excited. They can mimic the feeling of love. They can instil us with a sense of connection to everyone in the whole world. They can interact with brain regions that control imagination and produce vivid images and colours beyond one's wildest non-chemical imagination.

What they cannot do is everything at once. Because drugs masquerade as neurochemicals, they can only enter pathways designed for our internal chemistry. They can cause experiences by imitating those natural internal chemicals, by changing our response to that chemistry, or by altering the chemicals that umpired these effects. They cannot produce any effect that the body is not already predisposed to experiencing by its basic chemistry.

We experience things in our brains when neurons, the communicating cells of the brain, whisper to each other through a chemical signal sent across a small gap (synapse). The neuron sending the signal is the presynaptic neuron. The one receiving the signal is the postsynaptic neuron. The sender (presynaptic neuron) transmits the signal as a chemical message (neurotransmitter) across the synapse to another kind of chemical (receptors) on the surface of the receiver (postsynaptic neuron). When enough receptors (on the postsynaptic neuron) get enough neurotransmitter, the postsynaptic neuron receives the message.

That neuron can then play postman. It becomes a sender, and it passes the message to other neurons. In this way, neurons send messages around the brain, and to and from the rest of the body.

Once the presynaptic neuron sends the message, it also deploys the chemical equivalent of a dustpan (an enzyme) that wipes away excess neurotransmitter. The dustpan brings the pieces of the neurotransmitter back to the presynaptic neuron for recycling.

When a foreign substance gets into the brain, it may do nothing noticeable, or it may interfere with this signalling process. If it mimes a neurotransmitter, then the postsynaptic neurons will believe that the presynaptic neuron has sent a much stronger message. The interfering agent can damage the dustpans (enzymes) that sweep away the remnants of old signals, thus making it difficult to remove neurotransmitter from the synapse. It also can make it easier or harder for the receptors to respond to the message by increasing or decreasing the number of receptors (docking ports) on the receiving cell. In short, drugs use a range of mechanisms to send or alter their messages within the nervous system.

One very efficient way that drugs can affect the brain is to mime any of the chemicals known as neuroregulators or neurohormones. The brain releases these chemicals into specific regions rather than into specific synapses. Neuroregulators make the neurons in those affected regions more or less susceptible to neurotransmission. By altering an entire brain region, they can change the relative perception of that region. If that region happens to be a control center for vision, pain, or emotions, then the drug will alter that specific function.

The Chemical Lie of Opiates

When it comes to explaining how drugs work, few things beat opiates. Opiates include diverse pain-deadening medicines such as morphine, heroin, methadone, fentanyl, and opium. We first made opiates as derivatives of opium found in poppies, and later through synthetic variations of these molecules. They are very odd drugs because many people who try using them as recreational drugs have a negative first experience with them. The initial response to these drugs can include nausea, vomiting, and constipation.

Despite nausea, which one might consider a strong deterrent, people will actively seek the drug because of what comes next: euphoria and release from pain. I have heard users describe the feeling that they get from heroin as being better than the best orgasm of their lives. They feel a flush of well being, of total relaxation, and a melting away of all pain and cares. Anyone who ever just wished a pain would end, be it physical or psychological, can understand the allure of heroin and opium.

This euphoric sensation is the deception perpetrated by the opiate. When a person takes an opiate, the drug convinces the body that it belongs there. It mimes a class of our body chemicals, endorphins or endogenous morphines, which look very similar to the opiate found in poppies. Endorphins regulate a range of reactions in the brain, including pleasure and pain pathways.

The body makes endorphins when we consume spicy foods (500), when we experience love (501; 502), or when we engage in pro-social behaviours such as friendship (503; 504).

We also produce endorphins when we need to overcome pain (505; 506). On a daily basis, we must overcome the slight pains caused by the texture of our clothing, the discomfort of sitting too long, the muscular pain created by physical labour, and even the slight pain caused by chewing our food and breathing. Most people remain unaware of these pains because their endorphins block them.

Most of us can overcome these moment-to-moment variations in our pain receptors because our brains pump just enough endorphin into the system to mask them. People who exercise regularly and who then lose access to exercise for a few days can experience a mild exercise-withdrawal. As their bodies adjust to receiving less of a pounding at the gym or on the track, their brains produce fewer endorphins. As a result, these people crave exercise to raise their endorphin level.

The world contains another group of people too. These other people have a physiology that leaves them feeling wrong. No matter what, they never quite feel right, as if they had a continuous low-grade pain. Our perception of pain includes not just the external pain source and intensity, but also the interaction of the transmitters, the neuroregulators and the receptors for both.

Although a person cannot experience the pain of another person, we assume universality in pain experiences. That too is a ruse. Our species experiences a range of pain tolerance and pain severity, just as we experience different vision or hearing. Most people experience daily pain as a fleeting, temporary, and usually not very severe. Someone whose life or whose biology causes them to feel brutal chronic pain will seek relief. The person who experiences ongoing discomfort will seek to feel normal.

When we consume an excess of heroin (or any opiate), our brain becomes amazed and enamoured at its prowess. It thinks it made all those neuroregulators by itself. It does not distinguish between the natural endorphins and the fake ones that we have imported.

Because the euphoric effect interferes with its ability to function, the brain tries to restore balance through a process known as down-regulation. In down-regulation, the brain tries to compensate for what it perceives to be its mistake. It reduces the amount of endorphin that it produces, and it reduces the number of endorphin-receptors on the neurons. As the brain changes these internal controls, it needs more and more heroin to stimulate the brain into the pleasure response. As the body reduces its production of endorphins, it becomes unable to overcome even slight pains. Between these combined effects, the user soon ceases to take in the drug to feel euphoric. It becomes necessary to take the drug just to feel normal.

By the time the user has become aware of the drug's mimicry, it is usually too late. The body has lost much of its natural mechanism for pain and pleasure control. It has become physically dependent upon the drug. Withdrawal from the drug feels like the worst case of the flu one has ever had because the person is highly aware of every part of his or her body. The few remaining endorphins and receptors cannot screen the pain.

In most people, the brain figures out what happened, and begins to rebuild the pain and pleasure control systems within about five days. The brain begins to up-regulate the production of endorphins and endorphin receptors, making more and more to compensate for the missing drug. Within a few days to a few weeks, the endorphin system has mostly recovered. Some people will continue to experience low-grade pain for months or even for years, feeling their clothing as it chafes against their skin, or the pressure of their body against a chair when they sit.

Even though the pain fades, the memory of the euphoria, the romance of the lie, can remain much longer. Sometimes, it remains forever. This memory makes the person want to return to their romanticized relationship with the drug, just as so many people return to an abusive spouse. "Sure, there were problems, but it was not all bad," we think. When we tell ourselves that lie, we set the stage for *Act Two: The Relapse.*

The Chemical Lie of Cocaine

Our bodies contain an array of chemicals that manipulate our feelings and lead us to feeling happy, sad, angry, or afraid. Most classes of psychoactive drugs tap into these internal systems and mime these neurotransmitters and regulators. They then work on principles similar to the one that makes heroin attractive.

Cocaine mimes chemicals that make us think we are more powerful and dominant than we are. For a brief while, we experience a sense that mimes the confidence and sense of self-worth of someone in a far higher societal position. It couples this powerful feeling with increased alertness, a state many of us crave in our over-worked, overactive society. The cocaine-high does not just feel good, it feels wonderful—at least for a short while.

When taken, cocaine causes our body to drain its stock of the neurotransmitters made from the amino acid tyrosine. These transmitters give us the extreme high. Once the drug depletes our brain of these transmitters, all that remains seems hollow. From the heights of the lie, it then lets us down to a sense of greater despair than the one that initially drove us to the drug.

The behaviour of cocaine in the body seems more sinister than many other drugs. I have worked with many people hooked on crack, a particularly fast-acting form of cocaine. Time after time, I have heard people say the same thing. He or she decided to try it just once—often in curiosity, to try to understand why so many people loved it. Then, they were hooked. The addiction seemed almost instantaneous. Unlike so many other chemical dependencies, which slowly sneak up on people, cocaine forms a sudden and well-camouflaged pitfall-trap for many people.

We still do not fully understand why some drugs, like cocaine, have the potential for rapid addiction, while others take time to develop. One piece of the puzzle rests in the many chemical pathways on which cocaine works. It works on both the serotonergic or resting pathway—think heroin, alcohol and ice cream—and also on the catecholamine or alertness pathway—think dopamine, adrenaline and a wild party with friends. By combining these chemical paths, it acts as both an antidepressant and as a stimulant.

Another possible piece of the puzzle may lie in cocaine's ability to turn off genes that regulate aspects of brain chemistry. Cocaine induces specific changes in the proteins that control the output of messages from DNA or histones in the dorsal striatum, a brain region better known as a "reward center." This part of the brain is an important region upon which cocaine exerts its influence (507). The ventral striatum, also a reward center, is also impacted by cocaine (508). In rats, these changes begin within half an hour of a single dose of injected cocaine. The dorsal striatum of the brain signals a reward for using cocaine, making its use more and more attractive. A rapid increase in the chemical reward-signal, dopamine, is intimately involved in craving and dependency on cocaine (509).

Cocaine does not perform its feat by magic. It uses a well-established mechanism called methylation. Methylation and its sister, acetylation, allow genes in our body to switch on and off when we need to control them. Our bodies use this process throughout our lives. Some changes are permanent; our bodies can reverse others.

During development, we methylate and acetylate genes to turn one lineage of cells into heart cells, and another line of cells into our big toe. Cocaine is just one of the many things in the environment that can continue this process throughout the life cycle. Two identical twins may thus start life with identical—or at least nearly identical—DNA codes, but how that code expresses itself begins to diverge from the point of conception forward. Every experience in our lives, including our addictions and our desires, has the potential to alter this pattern of methylated genes.

Biologists have barely begun to explore this regulating layer of cellular metabolism that lies above the genome, the epigenome. In the future, it may hold a key to understanding and controlling the agonies of chemical dependency (510). It may help us understand why some drugs addict more quickly than others, and why some people seem more at risk for dependency. It may also provide road posts and warning signs for new kinds of antidepressants, such as the triple uptake inhibitors (511).

The Chemical Lie of Caffeine (and that Stuff in Tea and Chocolate)

Caffeine[46] is probably the most popular drug in the United States, Canada, and Western Europe. Drinks that contain caffeine include coffee, colas, yerba mate (chimarrão), guarana, and energy drinks. Caffeine is part of a chemical family known as the methylxanthines. This group also contains two other important sisters: theophylline found in tea, and theobromine, that adored stuff that occurs in chocolate. All

---

[46] "Caffeine" literally means "coffee stuff" and "theophylline" means "tea stuff." "Theobromine," the comparable compound in chocolate, translates aptly as "food of the Gods."

work in the body in similar ways, and together they form the most commonly consumed class of psychoactive drugs in the world.

Caffeine tells a lie that looks a little like cocaine's lie. It affects the striatum, like cocaine. It makes us feel elated and awake. Like cocaine, it stimulates dopamine usage in the brain, but it takes a different pathway to that end. Caffeine also triggers insulin release, making it easier for the body to access sugars in the bloodstream (512), and it increases the release of adrenalin into the blood. Like cocaine, caffeine increases the release of dopamine in the brain's pleasure centre, a trick that creates a reward and increases the desire for the drug

At the molecular level, caffeine and other methylxanthines look different from cocaine. The shape of caffeine makes it fit differently into the proteins of the brain than cocaine. Rather than miming tyrosine-based transmitters as cocaine does, a moderate dose of caffeine mimes the nucleic acid adenosine (513).

In the body, adenosine plays many roles. One finds adenosine in DNA. It functions in the standard energy currency of the cells (ATP). It also acts as a neurotransmitter. As a neurotransmitter, adenosine works very differently than its caffeine mimic. It makes you sleepy. It tells you to reduce your activity. Caffeine stimulates us because it interferes with the function of adenosine. When caffeine is in the body, the receptors cannot attach to enough adenosine, and the signal for rest goes unsent. By reducing our sensitivity to adenosine, caffeine and her sisters make us feel more alert. Consequently, we may ignore our body's need for rest.

The Chemical Lies of Other Drugs

As for other drugs, each psychoactive drug will interfere with a normal bodily process to cause its effect. Some drugs, like heroin, mime a hormone or a transmitter. Other drugs exaggerate the activity of a hormone or a transmitter. Still other drugs block receptors or disable the chemicals that attach to them. They can increase or decrease one's sensitivity to the internal chemical environment, or they can enhance or cloud the body's perception of its chemistry. In these ways, substances alter emotional, cognitive, or physical experiences.

The amphetamine ecstasy (MDMA) mimes love, empathy, and intimacy—feelings that most people crave. In the body, oxytocin triggers these emotions. In the early days of MDMA, some counsellors used ecstasy in couples' counselling to rekindle a sense of attraction between individuals. The artificial introduction of the feelings by the drug seemed to reawaken real feelings.

LSD (acid) and other hallucinogenics stimulate many regions of the brain and many receptors. This variety leads to experiences of odd things. We may see bright colours where there are none. The drug may tell us that we can do things we cannot do, such as floating through the air. We may taste colour or see a smell. LSD can also increase our sense of empathy and well-being. As LSD whispers to us, we may believe we are capable of creating great art, even though we might not even know how to hold a paintbrush or sculpting chisel.

Like heroin, these drugs cause biological changes in the brain and body as the body tries to compensate for the chemical lie. It adjusts neurotransmitters and neuroregulators downward, so the brain has fewer of them bouncing around its synapses. To accomplish these ends, it turns on and off different bits of DNA. Some of these changes in DNA may be short term; at other times, the damage lingers for a long time or may even be permanent.

By the time the drug has finished making these tweaks in brain function, the little lie told on that first encounter sounds sweeter than ever. Even years after the last fix, the user of some drugs, such as heroin and tobacco, often listen to the Siren's song. They will ask her to lie to them again, just for a little while. If we recall that these drugs are making use of the body's signalling system, it becomes easier to understand how non-chemical activities, such as running, gambling, or sex can also feel addictive.

The Chemical Lie of Love: Our Partners Feel As We Do

Unobtrusive, quiet and retiring, without being shy, humble and homely in its deportment and habits, sober and unpretending in its dress, while still neat and graceful, the dunnock exhibits a pattern which many of a higher grade might imitate, with advantage to themselves and benefit to others through an improved example.

~ Rev. Frederick Morris, cited by Nicholas B. Davies in *Dunnock Behaviour and Social Evolution*

In 1853, the Reverend Frederick Morris delivered the above sentence in a sermon to his parishioners (514). He told them to emulate the behaviour of the dunnock, a type of hedge sparrow. At that time, everyone *knew* that the dunnock lived the life of a proper, moral Victorian. Dunnocks were monogamous, never cheating on each other, and never straying from their faithful and loving relationships. One hundred years later Dr. Nick Davies discovered that the female dunnock often had multiple male partners. Like many *socially* monogamous birds, the dunnocks had one *social* partner but many *sexual* partners.

That does not mean that monogamy does not exist in the animal world. Had the Reverend Morris entreated his congregation to behave like prairie voles (the small rodent from Chapter 3), it would have been a more apropos example (515). Female prairie voles will drive away other males after they mate. Male prairie voles lose interest in other female voles after they mate. The American bald eagle and several species of goose also genuinely appear to be monogamous, changing mates or "divorcing" only after evidence of infertility or the physical loss of a mate.

Our species, and especially Western culture, puts much stock on love in relationships. There are cultural variations in the closeness and demanded fidelity in these relationships. In some cultures, divorce is simple, and in other cultures, it is complex. Sometimes a man will have more than one wife, and in a few cases, a woman will have more than one husband.[47]

The human pattern varies greatly. One common theme, however, is that males *tend* to guard their females much more openly than females guard their males. In a few human societies, female mate-guarding does become openly violent, but human females retaliate against a mate-stealer with lethal force far less often than do males.

In general, females will aggress, but they usually do so differently. Rather than using physical violence, they aim at relationships and reputations. Through this kind of bullying, they can reduce the status and the desirability of their competition (516; 517; 518).

---

[47] As I noted in Chapter 7 on the section of sexual sneaks, this latter pattern is usually more a matter of male relatives sharing a wife than a wife choosing multiple husbands.

Both men and women in our society assume much about the other gender. North Americans often take for granted that the other gender feels the same things that they feel. Both males and females spend much of their lives assuming a great untruth: that there are no significant psychological differences between them.

We behave as if the divide that separates us is in our imagination. We assume that differences between men and women come from gender-biased socializations rather than a combination of socialization and biology. Because we believe that different implies better or worse, we even have an understood taboo about mentioning the possibility of innate differences among the genders.

We are then dumbfounded by the behaviour of a friend or lover of another gender when he or she acts differently than we do. We see that person as unusual; we all but refuse to entertain the possibility that his or her gender perceives life differently.

Western society reinforces this lie with its way of thinking. Because they have historically held subservient roles in our culture, females are acutely aware of gender issues and want men to see them as equal, full, complete counterparts. Because males have enjoyed a position of status, many men are less aware of the issues of gender inequality than are women. To maintain the ignorance on both sides of this divide, it takes lies of omission.

North American society also views the world in hierarchical and competitive terms (519 pp. 79-80, 169). From birth, Anglo-North-American culture conditions us to see things in a hyper-competitive light. The Western mind is accustomed to seeing things either as equal to each other or as one being superior to the other one.

The idea that two things may not compare with each other seems almost incomprehensible. In our Western minds, confrontations should have a winner and a loser. We have best friends, and we have favourite ice-cream flavours. Companies often compete not to make products of excellence, but for the nominally best product or the largest market share (520; 521). Too often, we see disagreements in politics as battles between good and evil rather than between honest differences in ideas about how to serve a common interest.

This cultural lie is a simple one, but a seductive one: there can only be one right way of doing something or thinking about something. There must be a winner and a loser in every contest. In reality, the more options one has, the more secondary and tertiary ways we have to solve problems, the more flexible and more successful we become.

The hyper-competitiveness of North-American society bleeds into our understanding of gender. If our gender tends to act differently from the other gender, then one of us must have the better method. The other gender's mode of behaviour must be inferior. Our competitive, either-or way of thinking tells us that either we are the same, or one of us must be better than the other one. Women try to feminize men. Like Henry Higgins in *My Fair Lady*, (522) men wonder why women cannot be more like men.

We seldom stop to consider another possibility. Some things cannot be compared. They form a lattice rather than a ladder. Males and females have coevolved their behaviours, but those behaviours do not have the same function. Males and females respond differently to the presentation of visual sexual stimuli (523; 524). Women respond differently during different phases of their monthly cycle (525; 526). Men respond more strongly to the paedomorphic femininity of women's faces (a fertility marker) when their testosterone levels are high (527). Like parallel, non-intersecting universes, we may be unable to find a one-to-one correspondence between all aspects of gender behaviour.

When it comes to cross-gender communication, men and women assume they use the same system of communication. When a woman ends a date by saying "I have many things I need to do tonight," she means, "no sex." In her mind, she has communicated that the date has been fun, but she has no desire to let him become the potential father of her children. Too often, to the male ear, she has said something different. He hears a much more utilitarian message, "hurry up with the sex, I am in a rush." He thinks she has signalled for quick sex. She thinks she has said, "no." She ends up thinking he is a sex fiend, and he comes away from this encounter thinking she is a tease.

Although most of these miscommunications are not so much a lie as a different way of listening to the world, the results are often the same. Millions of years of evolution account for these differences in interpretation. Males receive a tangible reward for having sex: leaving potential babies behind them. Avoiding concurrent multiple encounters with multiple females (monogamy) can benefit a male if it tips the odds in favour of having more access to eggs and if monogamy helps more of his offspring survive to adulthood; otherwise, serial monogamy or polygyny increase his procreant odds (528; 529). Across species when there is a shortage of female partners, or when females demand time and resources before copulation, then monogamy becomes a more viable strategy.

Females receive several rewards for not having sex. First, they can delay pregnancy until great genes come around. Second, venereal diseases often impact women more than men because sexual fluids typically stay in the female longer than they stay in contact with the male. Penetrative sex also results in more damage to the genital mucosa of a woman than it does to the mucosa of a man, further increasing her risk of contracting any disease he might harbour (530; 531). The human female, or any mammalian female, also deposits less potentially infectious fluids on her mammalian male partner (530).[48]

Such differences in reproductive risks and rewards make pair bonding unusual among mammals in general, not just humans. Even in species where a social bonding has arisen to care for the young, females may seek out extra-pair copulations, if such a fling allows her to get better genes than those provided by her partner.

As we saw in Chapter 8, some species of "monogamous" birds are not monogamous. Many males may father a majority of the eggs in a nest. Up to 90% of the nests will show DNA evidence of female infidelity (532; 533). For those female birds, her social partner may be wonderful for building a nest and gathering food, but that does not mean he is the best sperm-donor for her eggs. Among mammals, a group in which the investment of the female is very high, we find that that social monogamy is extremely rare, with only about 3% of species displaying some degree of sexual monogamy (534).

---

[48] These differences, combined with the hidden nature of STDS in women, make the actual rate of STDs in women hard to evaluate. Men normally become aware of the problem as visible lesions develop. Women often have internal lesions rather than external ones, and thus have fewer indications of infection.

Lies of Devotion

Part of our misunderstanding of monogamy, and what it means to males and females, comes from how males and females become monogamous. The chemical pathways differ.

In Chapter 3, we explored this chemistry of monogamy through the changes in the brain chemistry of prairie voles. The brains of monogamous male mammals, such as the prairie vole, are loaded with receptors for vasopressin (V1-a receptors) that addict them to their partners. All monogamous mammals studied to date have brains riddled with these receptors. For the prairie vole, in the best studied of these monogamous systems everything really does change after sex (535; 536; 537).

Before mating, the prairie voles are not aggressive toward each other. Their oxytocin levels program these voles to be relatively passive and social. Oxytocin is the same hormone that induces rhythmic contractions in the uterus and testicles during sex. Physical contact, such as hugs in humans, stimulates its production. It makes us like the people around us and makes us feel more generous toward them (538). It aids in bonding between mothers and children and is an integral part of the female and male sexual orgasmic response. Oxytocin is the hormone that makes us feel safe, secure, and relaxed. Before mating, the presence of another vole pleases both the male and the female because it triggers the release of this hormone in their brains.

In these voles, smell triggers the interest in sex and the chemical changes that accompany it. Both the male and the female experience the smell of their mate as an addictive aphrodisiac. The molecules enter their little bodies through a specialized receptor in the nose used to detect pheromones—the vomeronasal organ. Once these sexual scents attach to receptors here, the differences between the male and female chemistry begin to diverge quickly.

The Male Vole in Love

In the male, the scent triggers the release of a compound, vasopressin, which physically stimulates him. Vasopressin resembles its chemical cousin, oxytocin, but these hormones trigger pleasure responses in different parts of the brain.

Male prairie voles predictably respond to these chemicals. They become more active and impassioned about the one with whom they are mating. As their oxytocin levels also rise, they feel a drop in the levels of stress hormones. In other words, they become alert and relaxed. He is getting to know her. Following his very first sexual experience, the male brain produces vasopressin, and a male prairie vole prefers to stay forever with the now love of his life (539).

The male becomes even less interested in other voles and other sex partners. Vasopressin reduces his interest in novel experiences, including sexual novelty. He becomes more interested in the security of the known partner and their nest. This hormone also binds to areas of the brain that cause the male to become aggressive, defensive, and territorial. Until their first mating and this chemical change, male prairie voles are not particularly aggressive toward other males.

The chemistry of the orgasmic response causes him to become "addicted" to his female partner, inducing monogamy. The regions of the brain stimulated by the intercourse include the areas stimulated by cocaine and amphetamines, a.k.a., uppers.

Non-monogamous males, such as montane voles, have fewer of these receptors and different genetic versions of them. Without them, a social attachment to the female does not develop. They do not get the strong neurochemical reinforcement. Montane male voles are quite content to love 'em and leave 'em.[49]

The Female Vole in Love

In female voles, the love-change also begins with the musky smell of the male. The musk and social interactions also trigger a corresponding rise in oxytocin levels in the female. These levels trigger changes in other sex hormones that cause her uterus to enlarge and to get ready for pregnancy (540). Her ovaries then produce fertile eggs within two days of their first encounter (541). The couple will continue to gain more and more interest in each other until, at just the right moment of fertility, the pair will mate—virtually non-stop—for a full day. Both of them will have had a wonderful, wild, hormonal roller-coaster ride. She has become addicted to him, and he has become addicted to her, but through very different systems.

During this process, the female does not seem to develop a rise in vasopressin like that seen in the male. Her response, her sense of love, is regulated predominantly by oxytocin. His sense of love is a combination of the conflicting messages sent by oxytocin and vasopressin. Both the male and the female prairie vole follow an

---

[49] After learning about this connection in class, many of my female students want to develop a saliva test to measure probable monogamy based on this chemistry. I need to remind them that the same chemical pathways induce mate-guarding (jealousy) in these species, and that the same pathway could be linked to other novelty-avoidance patterns, such as autism, that may not be good for their offspring.

addictive chemical pathway by following the rewards of their first sexual encounter. For the prairie vole, this addiction is severe. For the montane vole, the chemical reactions are not as strong, and an addiction to the sex partner never occurs. Although male monogamy is rare in mammals, when it does occur, one usually finds this same underlying hormonal system of oxytocin and vasopressin.

"One Man Woman/One Woman Man"

~ song by Paul Anka

Humans come with varying degrees of monogamous tendencies. The same system of hormones found in prairie voles controls these same monogamous tendencies in humans. Male humans who have high levels of vasopressin and oxytocin receptors tend to associate the rewards of sex with a particular partner or soulmate. They tend to be relatively monogamous. High vasopressin levels also make males perceive other people as less friendly, just like in the prairie voles (542). This perception, driven by the hormone, has nothing to do with the actual friendliness of people. The chemical change in the brain makes people *seem* less friendly.

Male humans with fewer vasopressin receptors tend to be less monogamous. Low levels of oxytocin or oxytocin receptors in males and females correlate with social amnesia or an impaired ability to link social and sexual rewards with specific partners (543; 544; 545).

It is hard to estimate the natural rate of human female polyandry, the practice of taking multiple male mates. From the perspective of biology, women can benefit from some polyandry (546). They may gain resources for their young from multiple males by confusing paternity. They can gain access to different genes for their young. (Recall that a good social partner is not necessarily the best genetic partner.) Through multiple partners, women also reduce their risk of copulation exclusively with an infertile male (547). They may also gain a modicum of protection from their tristing partners.

In much of the world, controlling human males distort the sexuality females might express if they were free of male restrictions. Men frequently guard females as jealously as the male prairie vole guards his mate. Women often lack the right to refuse sex to their husbands, and they may have little or no practical outlet for extramarital sexual encounters. This pattern even holds in societies that consider sex essential to the mental and physical health of both men and women (548; 549). Men typically have fewer social restrictions on their sexuality than do women. These complications make the comparative study of human male and female sexuality difficult. One should read the following with this proviso in mind.

Across many different measures and cultures, men seem to win the libido contest easily. Men report more frequent thoughts about sex and sexual fantasies, report a greater desire for frequent sexual intercourse, have more partners, masturbate more, and want more variety in partners (550).

The higher libido of males is not an artifact of a single culture. Males regularly report a higher number of partners than do females (551). Numerous measures across cultural boundaries demonstrate the higher sexual interest of men. Studies in cultures which consider the male desire for sex to be the driving force and in which women have little choice suggest that these couples have sexual encounters an average of two times per night (552). By comparison, adult Americans, where females are culturally permitted to refuse sex, reported that on average, heterosexual adults reported engaging in intercourse only 0.15 times per day or 57 times per year (553).

A Turkish study reported that the rate of premarital sexual contacts among men was 61.2%, but it was only 18% among women (554). Despite locker room brags about sexual conquests, adolescent American males typically also tend to have moderate sexual activity. A Korean study that interviewed unmarried 19-30-year-olds had 67.3% of men reporting sexual activity during the previous year. In that same study, only 30.0% of women in the same age range reported having had premarital sex (555).

That same study reported that 50% of the men had had their first sexual encounter by age 19-20. It took until age 27 for women to experience their first sexual encounter. Young men who are sexually active tend to have only about two partners in any given year and sex about 2.7 times in a month (556). Lesbians tended to engage in sex later than heterosexual men, but sooner than heterosexual women. Two-thirds had their first encounter before the age of 20. Often this first experience was with socially endorsed, male partner (557).

If you are like me, you are probably wondering, right about now, how heterosexual men can have so many more sexual contacts and experiences than heterosexual women. One would think it should take one of each to engage in heterosexual sex. Part of this discrepancy may be due to women under-reporting contacts or men over-reporting their conquests. These equations often leave out another factor: the range of contacts.

More men have multiple partners than do women. However, when women have them, they have many more. Specifically, in the web of human sexual contacts and connections, most men have more contacts than most women, but the women with the most contacts, especially sex workers, have far more contacts than most men (558).

The disparity between the male and female contact numbers also occur because researchers often discount rape, or they exclude or underrepresent prostitutes in studies of sexual behaviour (559) and because men often under-report how many contacts they have had with prostitutes (558).

Same-sex couples highlight these differences between genders. These couples tend to be more in accord with each other, despite the remaining societal pressures against such pairings. Freed from the limiting response of males, females form friendships that frequently continue after a sexual or social partnership ends. Freed from the limiting responses of females, males are free to explore more sexual options and have more sexual contact. Data from the early days of the HIV epidemic (before gay men became widely aware of the HIV risk) suggest that the mean number of male-male contacts among those men who regularly visited bars or baths, was about 7.5 (median 3) sexual contacts per three-month period. Even in the early days of the HIV epidemic, with the advent of HIV education, that figure dropped to 1.1% over a three-year study, with abstinence also increasing (560). Keep in mind that this figure is biased. Researchers based it upon one segment of the gay community, i.e., those who frequented bars and baths. Most gay men do not fit this description.

Despite its efficacy, we have continued to insist on denying adequate HIV education in our schools. The American Center for Disease Control reported that by 2014, only 24% of US public high schools provided some form of HIV-prevention education that addressed the concerns of sexual minorities (561). This low emphasis on vulnerable, at-risk sexual minorities correlates with lower condom use among young gay youth than among young straight youth (562). Because we prefer not to broach the uncomfortable subject of sexuality, we have instead chosen to lie to our ourselves and our children. As a consequence of our societal indifference, we continue to place young men at risk for HIV infections even after forty years of the epidemic.

In this respect, we parallel the willful disregard for life that happened when the United States forbade reports and education about the Spanish Flu because it might hurt the morale of the war effort (563 pp. 179-199) or the downplaying of the COVID-19 pandemic because it could harm the economy. Millions died then too, and these lies of omission helped spread deadly respiratory diseases around the world. For our lies of omission, we too must hang the heads of our generation in shame because of the many who died needlessly, and for the havoc that we have already caused to the future.

Chapter 12

The Social Cost of Lies and Our Daily Cheats

"Maybe even you and I might do it."
~ "Let's Do It" by Cole Porter

Lies may be adaptive for individuals, but for societies, lies among individuals can be costly. Social beings spend much energy trying to catch cheaters, those organisms that do not live up to their share of social contracts. Cheaters break codes of reciprocity when they take and take but fail to give. Human society makes exceptions for people who, because of age or a physical or mental handicap, either cannot participate or cannot understand the rules. We usually do not classify such people as cheaters, though we may readily victimize them.

The obvious examples of cheaters include the person who puts the cheap gift into the holiday grab bag or the neighbour who never takes a turn at ploughing the snow. Cheaters also include the people who stiff a server of his or her tip, who shoplift, steal, rape or use questionable deductions to avoid paying his or her share of taxes. (Honestly, your dog cannot be considered a "consultant," no matter how much you trust her judgement.) All of us engage in deceptive activities, but the ones we consider acceptable and the ones we consider despicable vary from person to person and from situation to situation.

Unfortunately, cheating is natural. We all engage in cheating, and we have basic psychological mechanisms to rationalize our behaviour. To protect ourselves from recognizing our social cheats, we do everything from denying that the naughty behaviour ever happened to rationalizing or to justifying it because of background events or societal prejudice.

The Importance of Lying to Oneself

> *"Roger Chillingworth was a striking evidence of man's faculty of transforming himself into a Devil, if he will only, for a reasonable space of time, undertake a Devil's office. This unhappy person had effected such a transformation by devoting himself, for seven years, to the constant analysis of a heart full of torture, and deriving his enjoyment thence, and adding fuel to those fiery tortures which he analyzed and gloated over."*
>
> ~ *The Scarlet Letter* by Nathaniel Hawthorne

Using our brains to deny the harms done to us, those harms that could destroy us, can be our salvation. Anna Freud elaborated on her father's work and described these "defence mechanisms" (564). In the late 20th century, physiologists identified hormonal correlates (565; 566; 567; 568) and neural regions and lateralization associated with some of her defence mechanisms, including rationalization (569; 570; 571).

Popular psychology often condemns these defences. It urges people to remember and to confront their past wounds. Pseudopsychology demonizes denial and suppression rather than treating these brain functions as one of the necessary, natural paths to healing wounds.

Placing past horrors in perspective can be therapeutic *if* paired with work toward a resolution of a problem, but the defence mechanisms, through which we suppress counterproductive thoughts and information, can break our fixation on traumatic events and protect us from disabling pain (572). Constantly reliving and strengthening traumas and resentment strengthens these memories, building them into our current behaviours.

Retrieval suppression of unwanted memories weakens their influence on our actions. The ability to control the intrusion of unwanted memories has a physical basis. It stems from changes in the architecture and function of the hippocampus (memory access) and the higher functions of the neocortex (573).

These defences help us through severe threats to our survival and to the more mundane task of rationalizing, minimizing, or obliterating our bad behaviours. Soldiers in Nazi Germany might very well have faced death for disobeying orders to annihilate Jews. To preserve their own lives and their sanity, many of them rationalized that Jews were not human or that they posed a real threat to the country, as portrayed in Nazi propaganda.

We rationalize taking advantage of others to gain resources for ourselves and our offspring. We choose not to think about the child, slave, or prison labour that contributes to the products that we buy. The brain of the heroin addict, whose brain believes it will die without the drug, will rationalize a relapse to save itself. We use defences to feel better about ourselves when we cheat on our diet[50], when we cheat on our spouse, or when we take a less-than-reputable path of least resistance to satisfy our wants.

---

[50] It just so happens that diet cheats rank among my superpowers. I can convince myself that calories consumed while working do not count, and that once I break my diet, I have the right to break it for the rest of the day.

The need to conserve the energy from food has typically programmed organisms to spend the minimal amount of energy to get the maximum amount of benefit (574). By taking this energetic path of least resistance, plants, animals, and even fungi and bacteria try to conserve calories and maximize reproductive potential.

Humans follow this same path. We choose where to place our efforts based upon our perception of rewards (575). We avoid studying for an examination until the last minute for the same reason healthy, able-bodied people still look for the parking spot closest to an entrance of a store: it costs us less energy in the short run. Sometimes these energy-conserving and cheating strategies work well, but if cheating spreads too widely in any social species, disaster can follow.

In colonies of shorebirds, such as gulls and terns, colony members will flock to attack threatening animals that come too close to their nests. The ruckus raised by the birds in most imminent danger will signal the surrounding birds to join in an attack (576). Together they dive bomb the predator, pecking at it, defecating on it and generally distracting it with loud noise and much activity. If some of the birds wish to raise an alert, they need to be convincing. A common perceived threat increases the wiliness of group members to participate in the defence of the colony (577). The good of the colony rests on their ability to convince the predator that eggs and baby-bird should not be on the menu.[51]

In each colony, some birds cheat. They fail to respond to the calls of the colony even after most of the other members join the fight. For some

---

[51] This same phenomenon can explain the differences in how Americans responded during World War II, the Vietnam War, and the Middle East War. The perceived differences in the threat to the society mirrors the wiliness of its members to go to war, a human behaviour that correlates with the risky mobbing behaviour of the birds.

physiological reason, the distress calls of their neighbours or the presence of danger does not trigger the same communal mobbing as it does in most of their roost mates. The cheaters will continue to sit on their nests, protecting their own eggs rather than protecting the colony. These birds reap the benefits of the colonial behaviour without having to expend the energy on the attack itself. This cheating behaviour can give them a slight reproductive edge over their fair-acting counterparts.

As long as enough of the colony continues to respond to distress calls, the cheaters can benefit from their inaction. If a colony has too many cheaters, the ability to mount an effective defence drops and the cheating strategy becomes ineffective. In a gull population consisting of all cheaters, everyone dies. In a population where everyone cooperates, cheating can be very advantageous, and it easily infiltrates the gullible population of cooperative birds (578).

There are points (Nash equilibria) at which this ratio of cheaters to non-cheaters becomes stable (579). In a Nash system, conditions may not be optimal for everyone involved. Participants, or players, make their choices based upon the behavioural choices of their neighbours. They work towards goals that might not be the ones they would choose if there were not competition or cooperation, but they behave in a way that maximizes their benefits given the goals and behaviours of their competition.

At a Nash equilibrium, the strategies of all parties become stable. Any behavioural shift will have less favourable results. In the flock, the cheaters may get to survive with the protection of the herd or flock. The honest birds get to reproduce at a high enough rate to remain in the population at a consistent proportion despite the presence of the cheaters.

In Western societies, the cost of seemingly minor cheating affects all of us. Shoplifting and employee theft do not only harm the store. If a store has only a 10% profit margin, it must sell $10 worth of merchandise just to offset every $1 in stolen goods. If a store has a 1% profit margin, they need to sell $100 worth of goods to cover that loss. Stores pass the cost of the losses and the cost of the security system to all the consumers.

The average retail shrinkage or theft rate is about 1.3% of inventory (580). If one assumes a 1.8 trillion dollar retail sales figure, approximately 31 billion dollars are lost to theft each year (581). This figure is a *low* estimate of this impact. Finklea (582) estimated that organized (gang) shoplifting alone accounts for $15 billion to $37 billion in annual losses and that 96% of retailers have suffered losses due to boosting. These data do not include the costs and dangers of relabeled and expired products

Fear of cheating can be as costly as cheating itself. Fear of home robbery has spawned a multibillion-dollar home-security business. American politicians' fear being called "soft on crime" or "soft" on terrorism has led to draconian penalties and a distaste for reducing precautions even when the threat levels are low (583). Billions of dollars that could be spent on everything from educational, social, or business programs to paying down national debt have become trapped in a system of punishment. Dollars get spent on preventing what we perceive as inevitable random crime rather than on actual threats or on efforts to reduce the root causes of the behaviour.

Deceit and Social Status

Hoke Colburn: "Oh, Miss Daisy, yesterday, while you was out visitin', I went and ate a can of your salmon. Now, I know you said eat the left-over pork chops, but they was kinda stiff. So, I stopped at the Piggly Wiggly and got you another can. You want me just to go on and put it in the cupboard?"
~ Driving Miss Daisy, directed by Bruce Beresford

In the film *Driving Miss Daisy* directed by Bruce Beresford, the elderly white Miss Daisy, who can no longer drive safely, objects to having a black chauffeur. In a classic scene, she announces triumphantly to her son that she has caught the driver stealing! The dramatic scene leads to her throwing open the pantry door to reveal the horrific theft: a can of food is missing. Unaware of what has just transpired, the chauffer enters with a paper bag, and he explains that he got hungry the previous day and ate the food. He then produces a can to replace it and puts it on the shelf.

Miss Daisy's world of racism and social caste lead her to expect all black people to lie. Around the world, people of higher social rank see casts of lower social rank as being less honest and more prone to crime than themselves.

Paradoxically, this tendency does not always go in two directions. Our species has a built-in program that pushes us toward obedience of perceived legitimate authorities. People of lower social status may resent the dominant primates, but they often work to help them retain their power. Our shared social constructs may bestow images of benevolence, wisdom, or holiness onto the powerful, and the underclasses may accept these images that often are imposed from above.

This distortion has its roots in the human tendency to construct paradigms that support our retention of power (584) and the dehumanization of our competitors to permit violence (585). We disproportionately punish blue-collar crimes as compared to white-collar crimes, even though waves of white-collar crimes have harm far more individuals (586; 587). The financial crisis of 2007-2008, for example, was catalyzed by the marketing of collateralized debt obligations and by people who were willing to take advantage of people with little financial means or knowledge. None of the major financial corporations or managers that fueled the crisis accrued serious harm or prison time (588).

Such distortions and falsehoods are critical tools for controlling people and retaining power for a group or within a group. We dehumanize adversaries in war, portraying them as intrinsic threats and our moral inferiors (589). We extend this behaviour to other areas in which we strive for power.

Politicians and political organizations often depict rivals as the bane of humankind. Even in the absence of war, the portrayal of an outsider in a negative light can be a useful political tool.[52] Too often, members of religious sects will claim moral superiority to other sects and non-believers. Some may not overtly claim superiority, but they may do so obliquely with references to "sectarian" values, be they Buddhist, Christian, Hindu, or Islamic, thus implying ownership of those pieces of morality. Others may tie pejorative terms to outsiders, such as unclean,

---

[52] Although not technically deceptive, it is at least as effective to take a group for which there is real evidence of unsavoury behaviours by the community or by their leaders, and increase animosity by repeating the charges (e.g., North Korea (666).

godless, or terrorist, while avoiding labelling the same or corresponding behaviours when done by a member of their group (590).

In war, the repeated, subliminal misrepresentation of a group can result in hatreds that allow their annihilation as vermin. In more social settings, the prejudices formed by dishonest and differential negative representations can result in hatred, social exclusion, or violent attacks (591; 592; 593). With a little sleight of hand, an empowered constituency can extend the cultivated fear of one minority group to engender hatred for another group, thus further coalescing its power (594).

Like Ms. Daisy, many white Americans have learned to see black Americans through a lens of fear and trepidation. Rather than realizing that African Americans are the disproportional victims of crime, Euro-Americans often characterize them as more violent and more prone to crime than their white counterparts. Rather than seeing African Americans as the primary victims of gang violence, European Americans often assume that violent gang behaviour is the norm in that community while dismissing similarly violent behaviours of white criminal gangs as isolated incidents. In Northern Ireland, Protestants tend to view Catholics through this same biased lens and as more prone to crime.

In Rwanda, the portrayal of the Tutsi minority as inferior and vile led to one of the more rapid and violent genocides of the 20th Century. Approximately 10% of the violence has been linked to anti-Tutsi broadcasts from specific radio stations. The distortions from these stations and its spillover into local rumours may have spurred as many as 51,000 people to hatred and genocide (595).

Part of our discriminatory vision comes from our innate tendency to divide the world into an I-Thou dichotomy. A group will ascribe positive traits to itself and negative traits to any group perceived as outsiders or as a threat. Our innate tendency to see people outside of our self-identified reference group as being homogenous enhances this dichotomy (596).

All peoples make this prejudicial distinction, painting their own in-group in a positive light. It becomes dangerous when this distinction goes unrecognized as a false dichotomy and when one group has power over another. The desire for power may not always be apparent. We do not always recognize the drive in ourselves, but the power of that quest predates our humanity.

Primates crave power (239). Power means you can do what you want to do. Your will and your way of doing things dominate. Males with power can gain access to more females and leave more offspring in their wake. You can impose your morals and your cultural values onto those with less power. No one below you in rank can do the same to you. No primate willingly surrenders that power without a struggle. By accumulating power—and its modern surrogate, wealth—one can increase the distance between groups and reduce the ability of the subjugated class to revolt. To this end, economic cheating can work very well.

A minority that experiences such discrimination still has the innate human desire for status. Where the haves may cheat to retain and to expand power, the have-nots may cheat to try to even the score. Humans who do not feel as though they have received adequate compensation for their contributions will try to balance the scales.

Slaves and indentured workers will not willingly exert more effort than necessary for the sake of their masters.

Employees who receive the lowest compensation, who are overworked, or who have the weakest connections to a company will tolerate or even condone theft from the company (597). Wage labourers who feel underpaid or under-appreciated will often reduce their output to a level they believe is reasonable for their level of compensation, or will take from their employers to make up the difference[53] (598; 597).

In contrast, people who feel respected and fairly compensated by their employers, or who have the opportunity for employee ownership, follow the predictable primate pattern of working with a fair alpha leader. They return the good intent and respect with an increase in productivity (revenue), effort, and/or loyalty, along with less wasted time and/or theft of resources (599; 600) (601).

As our society demands more and more from individual workers and treats them more and more as throwaway replacement parts, we should expect these phenomena to become more widespread (597). We will see less loyalty to companies as they display less loyalty to their workers. In response to poor treatment and cheating of the workers by companies, we can expect more retaliatory cheating of the company by employees. We can also expect a rise in stress disorders among workers.

Similarly, we can expect to see more protests against governments that the masses perceive as being exploitive of the rank and file or simply being out of touch with the suffering of their people. Although

[53] The pattern discussed here does not apply to theft and embezzlement among the higher echelons of power. There, a sense of entitlement can spur the theft rather than a sense of injustice.

they took very different forms, the Arab Spring movement in Lebanon, the Yellow Vest Movement of France, and the election of Donald Trump in the United States all had roots in populations who felt they were exploited.

As the status difference between workers and administrators increases, we should see complementary problems from administrations. As leaders lose contact with workers, the leaders are likely to show greater arrogance, entitlement, and executive greed (602). Wage theft will increase through the misclassification of labour and by use of freelancers to avoid paying benefits and taxes (603). Along with the rationalization of time and wage theft from employees, these conditions encourage companies to treat employees as disposable production units rather than as people (604; 605). We can expect the managerial positions to attract more people with antisocial tendencies as the rewards and the power increase.

As the status divide increases, we can expect to see the sense of entitlement grow among administrative personnel. We can expect to see a blossoming of corporate greed manifest in disproportionate grabs for riches by CEO's and other administrators. The sense of entitlement will also lead to more cheating of the company for a very different rationale. The communication and social ties with the actual units of labour—the workers—will be strangled through miscommunication, misunderstanding, assumptions of deceit on the part of the workforce, and the desire to maintain the power differential. The divide can result in two different classes of cheaters, each of which feels justified. In their eyes, they are compensating for being cheated by the other camp. Such feedback cycles of cheating and suspicion become increasingly hard to break as time goes by, and the impact on corporations from their joint effect can be deadly.

"Not a Brother, but a Traitor"

~ Gang Awareness Campaign, as quoted in a *Chicago Tribune* article "Evanston to Push Ads on Gangs" by B. Cox

Cheaters that benefit from crime do not always wear a simple face. Criminal organizations benefit from the poverty and despair generated from the experience of victimization. When an individual becomes a victim, he or she experiences neurological and biochemical changes in his or her body. The victim experiences these changes as acute- and post-traumatic stress. The physiology of trust changes and survivors have greater difficulty trusting and interacting with people over a wide range of daily activities (606). Repeated victimization leads to a feeling of learned helplessness that keeps the person from acting or from succeeding. The repeated experience of failure, even when induced by the actions of others, leads to emotional paralysis and an expectation of future failure.

For a criminal gang that makes its money from the despair of others, the key to success is to make others believe that they have no hope except through the leadership of the gang. The organization can then recruit new members with its favours, or it can sell limited glimpses of happiness through drugs, gambling, and sex. It can provide for the basic needs of its members and destroy them should they rebel. If it can destroy the paths away from poverty and despair, then a criminal organization can control its victims as efficiently as any Medieval fiefdom (607).

This human temptation to cheat by dominating others is a variation on a common theme among many social animals: raising one's status in society by reducing the status of others. It is not just the human schoolyard bullies who pick on the weak to elevate their social status. In colonies of shore-nesting gulls, the birds spend much of their energy punishing their neighbours. By pecking the birds in the next nest, they alter the neighbour's startle response. The peckers implant an understanding of the hierarchical relationships among the surrounding birds. The pecked bird will jump more easily, thus exerting more energy at protecting its nest and the nests of its neighbours from real or perceived threats. It also rewires the brain of the bullied bird, making it more susceptible to future stressors. In essence, the pecked bird will now jump when the pecker wants it to jump. Among gulls, it is better to peck than to be pecked. Social wasps will similarly enforce social position within a nest by exerting dominance both through posture and through stinging.

Closer to home, other mammals can also rule through intimidation. Aggressive hyena alpha-females signal their status with displays of a pseudopenis, a modified clitoris that looks like a penis (608; 609). The presence of the pseudopenis falsely signals to potential predators that she is a male—usually a less desirable target. Dominant females have access to the best hunting grounds in the heart of their territory; low-ranking females may need to hunt in more dangerous areas away from their territorial core (610). Dominant females signal their rank over other females with displays of the pseudopenis. This display results in subservient behaviour on the part of the other females, thus avoiding many fights. The relative social statuses of the females, imposed by the leader and reinforced by the submission of the others, tends to be passed to their daughters (611). Using the pseudopenis, although couched in deceit, remains an effective way to communicate among the hyena because they each understand what the signals mean.

Hamadryas baboons are socially similar to hyenas, except they have hyper-dominant males. Hamadryad males control harems of females through systematic harassment (369). Female hamadryads learn from infancy that the dominant males control their lives. Displays and behaviours intended to intimidate females by males is not unique to the aggressive hamadryads, but can be seen in many male primates, including humans (365; 612; 613; 614). Sometimes these displays pose real threats by males who are bigger and more powerful than the females; at other times, the threats are all hype. However, for the harassed and harangued primate females, who are typically smaller and less muscular than the males of their species, erring on the side of caution can be the wise course of action.

Remembering Our Other Appetites

Sex is not the only resource for which we are willing to fight and lie (615). Humans are environmental generalists. We eat just about anything from bugs to plants to fungi. We have conquered six of the seven continents and most larger islands, and we call them our homes. We live in houses, in trees, and underground. Our primary mode of adaptation to our environment and our culture rests on our ability to adapt behaviours. This flexibility also bestows an element of choice within the constraints of our biology.

Today, many people use our generalist behaviour to forage for information on the internet rather than for food on a savannah or sexual partners within our troops. However, we still graft our behaviour onto that pre-existing matrix of generalist behaviour (616; 617).

In the late Middle Ages, as books became abundant, misinformation also spread rapidly. Too often we still believe that if a book presents something as a truth, we must accept it as such. As we have transferred much of our first attention to print to electronic sources, our ability to spread misinformation has also become greater.

Our biology has not kept pace with our technology. We still must use our built-in, innate lie-generating and lie-detecting adaptations in this new social environment (618). When we hunger for information about a problem in our lives, we are likely to accept the answers that are most frequently cited or most easily accessed (619). We like frequently cited information because, like bees searching for a new home or a place to forage, we are programmed to trust the wisdom of the troop. We place great weight upon the frequency of a report as we analyse data. Our inbred use of optimum foraging to take the path of least resistance, i.e., our inbred laziness, steers us toward those answers that are easiest to access.[54] Consequently, if there are frequently cited, easy to access answers online, we are likely to accept them only because they are easy to find and many people seem to agree.

Once a popular, easily reinforced, false idea becomes established within a person or population, it can become entrenched. Once cemented into place, even large amounts of contrary information may not suffice to push it onto the trash heap of bad ideas. The false belief that childhood inoculations cause autism stands as a prime example of this pattern. The initial report of such a link was published in *The Lancet*

---

[54] Our desire for easy answers and easy access helps to explain why we so often fall victim to scams such as click-baiting (669).

(620) and included information that has since been discredited by multiple studies. The spread of the belief depended upon our human tendency to confuse correlation (two things happening close together) with causation (one thing causing another). Wakefield, an anti-vaccine activist, confused the recognition of autism at about the time that children received vaccinations with the idea that the vaccinations caused the condition. Within six years of the Wakefield article, vaccination rates in the UK against measles, mumps, and rubella dropped by over ten percent (621).

One of the core principles of science holds that given the methods, anyone should be able to replicate the results of a study. Because of the gravity of Wakefield's allegations, many researchers unsuccessfully tried to replicate the results of Wakefield's team. The inability to get the same results aroused suspicions about his research. Then, in 2004, the *London Times* disclosed a financial conflict of interest: Wakefield did not disclose that his research had been funded by a lawyer who sought evidence against vaccine manufacturers (622). Ten of his co-authors quickly abandoned Wakefield and retracted their support of its content, stating that there was no causal link between autism and the MMR vaccine (621).

The *British Medical Journal* described the initial article as a fraud (623; 624). *The Lancet* took the rare step of retracting the article on 2 February 2010. Other journals soon followed suit, retracting articles by Wakefield because of their questionable reliability or their dependence upon the original paper.

The damage done by the Wakefield report goes beyond mere non-disclosure. The damage goes beyond the charges of dishonesty and unethical behaviour toward their disabled patients.

Because of the very human fears aroused by this fraud, many parents became afraid of vaccinations. Misinformation on the internet continues to feed the anti-vaccination frenzy that lives on more than two decades later. Parents who want to know who or what to blame for the autism in their child have found a convenient and often repeated answer in this scapegoat. Unfortunately, by focusing on something that does not pose a real threat, we have inadvertently pulled needed resources from promising areas of research, especially prenatal research into the more probable causes of autism.

We have no idea how many children have died because they were not vaccinated, though several outbreaks in the United States and Canada have been attributed to the fears instilled by this paper (625). A 2014 report by the Center for Disease Control in Atlanta estimated that vaccinations will prevent 21 million hospitalizations and more than 734,000 deaths among children born between 1994 and 2013 (626). We do not know how many outbreaks of preventable childhood illnesses will kill or disable children before the fear can be dispelled. We do not know how many immune impaired people and older adults will die because of contact with this preventable illness.

In science, there is only one unforgivable sin, the purposeful fraudulent presentation of data. Someone caught in a lie, or who defends a result that they know to be incorrect, loses face and credibility in the eyes of their peers. Wakefield is not the only scientist to be ostracised by the scientific community because of fraud or because of his defence of discredited research. Each time some scientist breaches the ethics held by the scientific community, it erodes trust and can cost society untold amounts of time and research money.

In contrast, scientists will readily forgive honest errors, especially when the person who made the mistake fesses-up without prompting. In 1992, an astronomer named Andrew Lyne addressed a room packed with his collogues. They expected to hear him speak on his announced discovery of a planet orbiting a pulsar. Lyne and his team had published the paper on this unique discovery just a few months earlier (627).

What the audience did not know was that during the previous week, Lyne realized that he had incorrectly computed the impact of the movement of the Earth on his data. Once he accounted for the movements of our planet, his new planet disappeared. Andrew Lyne had to stand in front of an entire room of his colleagues and say those words that most of us dread, "I was wrong."

The response of the scientific community to his public admission of error was dramatic. The audience gave him a standing ovation (628). He did not try to mislead others to hoard grant monies for himself at the expense of his colleagues. He did not cross his fingers and hope that no one would catch his error. More important to the ideals and moral sensibilities of the community, his courage prevented the establishment of false ideas that could impact the research of others and the reputation of science as a whole. Andrew Lyne helped us to learn.

As we learn from our experiences, our choices become channelled, and fewer behavioural options become available to us. We learn to lie, and we learn how to detect lies. We may understand that what we have been told or what we may have experienced is not the entire truth, but we retain the adaptive biology of our ancestors.

We respond and adjust to the experiences around us. We live, love, and mate based upon a game of survival written millennia before we became modern humans. We cling to our belief that we hold our life-experiences in common and that our wisdom will thus work for others. As humans, we hold onto this belief even though we recognize and sometimes celebrate our differences in experience. We take advantage of each other at the same time that we object to others taking advantage of us.

One of our great strengths is our ability to recognize deceit in ourselves and others. When we take a step back from our daily activities, we can reflect upon all of our behavioural inclinations with our wonderful generalist brains. Without awareness, there is no hope for change.

Becoming aware of a behaviour does not necessarily mean we will control it, but it leads us in the right direction. The human brain, our great organ of behavioural adaptation, deceives us, helps us to deceive others, and gives us the tools to understand when others deceive us.

We do not have the power to reshape our nature by a pure force of will. Our DNA, experience, and physiology confine us within set behavioural boundaries. We do have the power to explore it and to discover its potential boundaries and to find out why a species that is so genetically limited can produce both the honest man and the corrupt cad.

Like all social animals, we must manoeuvre through this behavioural jungle. Unlike the lies of the orchid or the chimp, the potential consequences of human deceit in our modern era have consequences for everyone who shares this planet.

To all of our experiences, we bring both our fundamental ability to lie and our astonishing facility to overcome our programming for survival and deceit. We bring with us the adaptability to integrate these basic behaviours into our cultural behaviour. That ability is our greatest adaptation and our greatest hope.

Epilogue

"Just as the liar's punishment is, not in the least that he is not believed, but that he cannot believe any one else; so a guilty society can more easily be persuaded that any apparently innocent act is guilty than that any apparently guilty act is innocent."
~ *The Quintessence of Ibsenism* by George Bernard Shaw

Virtually every species depends upon lying and lie-detection as essential survival skills. We lie, exaggerate, and scam our ways to food, territory, and sex. Our brains lie to us, convincing us that our experiences are unique, or that our partners feel precisely the same way we feel. Our bodies lie to us, telling us we need food or drugs when these substances might harm us. We lie to become invisible or to increase our visibility. We pretend we are stronger than we are, or we pretend we mean no harm.

Duped individuals can become dinner rather than dates. The trickster may lead them to work against their best interest in favour of the interest of another. Through the muse of Loki, we become the biological fall guys for our fellow creatures.

We communicate honestly with each other when it serves our needs, but lying begins where common goals part company. For that reason, organisms have lived in an information-based arms race through most of our time on the planet. Good liars prey on the easily duped and thus select for the survival of the better lie-detectors. A good lie-detector can gain territory or food that might otherwise be taken. Through lie-detection, an organism can forge alliances with others and have a better chance of personal success.

By remembering that Loki lurks in the shadows, we can better assess potential mates and avoid the sneaky, creepy types that would impose their will upon us. We can evaluate the status of potential business partners and of politicians whom we may elect to run our lives. We can begin to understand how chemicals that alter our biology can create false impressions of the world and their benign nature.

Through awareness, we can make better judgements about those whom we choose to follow when we understand that a false aura of self-confidence and power, can cloud our perception of the most incompetent leaders who amount to little more than a pitiful psychopath, misleading us to obey them blindly. By understanding that lie-detection begins to fail in the face of too much disinformation, we can treat ourselves with a bit more fairness, making us less likely to suffer the ravages of our failures to detect all lies.

We can better understand the desires and intentions of potential sexual partners. When we respect our partners and to try to understand them, we can lose our cartoonish image of the other gender as an awkward, incompetent version of ourselves. Women open the potential of understanding why men act as they do. The full complexity of the male will overshadow the caricature of the sex-craved perpetrator. Men have the potential to understand why women crave security and friendship in their mates. This understanding eliminates the caricature of the gold-digger and replaces it with an understanding of the roots of behaviour.

As we learn to understand why children lie, their behaviour ceases to perplex us. It gives us the potential to be better parents when we understand that the goals of the child and the goals of the parent often differ. As parents, we can initiate positive interventions rather than simply responding to their perplexing behaviour. Understanding their powerlessness in our presence helps us to understand why children lie to us. Understanding our need to control helps us understand why we lie to children.

These lessons about deceit and power transfer directly to other uneven power structures. We can begin to understand different perspectives in our human problems with prejudice and discrimination.

Animals, plants, fungi, and the molecules of life all have the potential for deceit and trickery. Lying, the wisdom of Loki, helps us get through our lives, even if it costs us the unconditional trust of others. The trade has worked for billions of years. It will probably keep working for another billion.

Appendix A

"Mimicry" or "crypsis"

As the name implies, mimicry is a form of dishonest communication. It occurs when one organism (the *mimic*) attempts to make another organism (the *dupe*) think that the mimic is something other than what it is (the *model*). Both predators and prey partake in mimicry.

There are many types of mimicry. Mimicry can be defensive where the mimic tries to avoid an encounter. It can also be an offensive form, in which the animal uses the mimicry to get close to and to ambush the dupe. Animals can mimic other animals. They can mimic plants. They can mimic another prey or predator. Others will mimic inanimate objects. Mimics can feign behaviours or appearances. Sometimes several species mimic each other, with each one gaining an advantage by displaying a similar pattern. Some species have a part of their body mimic another part or have part of their body mimic a different animal or plant. Morphology, behaviours, and smell are all part of mimicry. Anything that can confuse predators, prey, or competitors can give an organism a selection advantage.

About the Author

Dr. Ptera Hunter, B.Sc., Ph.D., is a professor of Biology and has been a dean of studies at Noble International University. She received her Doctor of Philosophy in Zoology with a minor in Archaeobotany from the University of Toronto. For many years, she has taught classes on environmental biology, animal behaviour, evolutionary biology, parasitology, and animal diversity. She has previously published a children's book on insects, *The Unabashedly Totally Incomplete Book about Bugs*, several books of short stories and poetry, and several journal articles. Her writing is her avocation, but teaching is her life. In the spirit of the book and with the inspiration of Loki, she has chosen to use a pseudonym.

Citations

1. *The evolution of deception.* **Bond, C. F. and Robinson, M. R.** 4, 1988, Journal of Nonverbal Behavior, Vol. 12, pp. 295-307.

2. **Mitchell, R. W. and Thompson, N. S.** *Deception: Perspectives on Human and Nonhuman Deceit.* Albany : State University of New York Press, 1986. p. 418. https://www.amazon.com/Deception-Perspectives-Nonhuman-Deceit-Behavior/dp/0887061087/ref=sr_1_1?s=books&ie=UTF8&qid=1532879415&sr=1-1&keywords=Deception%2C+Perspectives+on+Human+and+Nonhuman+Deceit.. 978-0887061080.

3. *Animal behaviour: rank crime and punishment.* **Strassmann, J. E.** 2004, Nature, pp. 160-162.

4. **Gilbert, W. S. and Sullivan, A., [comps.].** The Pirates of Penzance (= The Slave of Duty). 1879.

5. **Mitchell, R. W.** A framework for discussing deception. [book auth.] R. W. Mitchell and N. S. Thompson. *Deception, Perspectives on Human and Nonhuman Deceit.* Albany : State University of New York Press, 1986, 1, p. 389.

6. *Regenerated claws dishonestly signal performance and fighting ability in the fiddler crab Uca mjoebergi.* **Lailvaux, S. P., Reaney, L. T. and Blackwell, P. R. Y.** 2, 2009, Functional Ecology, Vol. 32, pp. 359-366.

7. **Vasek, M. E.** Lying as a skill: the development of deception in children. 271-293. [book auth.] R. W. Mitchell and N. S. Thompson. *Deception: Perspectives on Human and Nonhuman Deceit.* Albany : State University of New York Press, 1986, 16, p. 418.

8. *Aggressive Chemical Mimicry by the Bolas Spider Mastophora hutchinsoni: Identification and Quantification of a Major Prey's Sex Pheromone Components in the Spider's Volatile Emissions.* **Gemeno, C., Yergan, K. V. and Haynes, K. F.** 5, 2000, Journal of Chemical Ecology, Vol. 26, pp. 1235-1243.

9. *Firefly femmes fatales: A case study in the semiotics of deception. Biosemiotics.* **El-Hani, C. N., Queiroz, J. and Stjernfelt, F.** 1, 2009, Biosemiotics, Vol. 3, pp. 33-55.

10. *Notes on the biology of some common Lampyridae.* **Hess, W. N.** 2, 1920, Biological Bulletin, Vol. 38, pp. 39-76.

11. *Aggressive mimicry in Photuris: Firefly Femmes Fatales.* **Lloyd, J. E.** 3684, 1965, Science, Vol. 149, pp. 635-654.

12. *Occurrence of aggressive mimicry in fireflies.* **Lloyd, J. E.** 3, 1984, Florida Entomologist, Vol. 67, pp. 368-376.

13. *Firefly semiosystematics and predation: a history.* **Lloyd, J. E.** 1, 1990, The Florida Entomologist, Vol. 73, pp. 51-66.

14. *Versatile Aggressive Mimicry of Cicadas by an Australian Predatory Katydid.* **Marshall, D. C. and Hill, K. B. R.** 1, 2009, PLoS ONE, Vol. 4, p. e4185.

15. **de Waal, F. B. M.** *Chimpanzee Politics: Power and Sex Among Apes.* New York : Harper and Row, 1982. B00NSYEAAG.

16. *Arms races between and within species.* **Dawkins, R. and Krebs, J. R.** 1161, 1979, Proceedings of the Royal Society B: Biological Sciences, Vol. 205, pp. 489-511.

17. **Dawkins, R. and Krebs, J. R.** Animal signals: information or manipulation. p. 282–309. [book auth.] J. R. Krebs and N. B. Davies. *Behavioural ecology: An evolutionary approach.* 1978, p. 512. https://www.amazon.com/Behavioural-Ecology-Evolutionary-Approach-Krebs/dp/0632002859/ref=sr_1_4?s=books&ie=UTF8&qid=1532888466&sr=1-4&keywords=Behavioural+Ecology%3A+An+Evolutionary+Approach..

18. *The evolution of large size: how does Cope's Rule work?* **Hone, D. W. and Benton, M. J.** 1, 2005, Trends in Ecology and Evolution, Vol. 20, pp. 4–6.

19. *Songbird cheaters pay a retaliation cost: Evidence for auditory conventional signals.* **Molles, L. E. and Vehrencamp, S. L.** 1480, 2001, Proceedings of the Royal Society B, Vol. 268, pp. 2013-2019.

20. **Lloyd, J. E.** Firefly Communication and Deception: 'Oh What a Tangled Web. p113-128. [book auth.] R. W. Mitchell and N. S. Thompson. *Deception: Perspectives on Human and Nonhuman Deceit.* Albany : State University of New York Press, 1986, 6. https://books.google.com/books?hl=en&lr=&id=eHV_8YC_NL0C&oi=fnd&pg=PA113&dq=Lloyd,+J.+E.+(1986).+Firefly+Communication+and+Deception:+%E2%80%98Oh+What+a+Tangled+Web.&ots=78vZVj9HfR&sig=EwbmvK4TnJw4FY7EzKARoW98A_Y#v=onepage&q&f=false .

21. *Machiavellian intelligence.* **Byrne, R.W.** 5, 1996, Evolutionary Anthropology, Vol. 5, pp. 172-180.

22. **Maestripieri, D.** *Macachiavellian Intelligence: How Rhesus Macaques and Humans Have Conquered the World.* Chicago : University of Chicago Press, 2007. p. 192. https://www.amazon.com/Macachiavellian-Intelligence-Rhesus-Macaques-

Conquered/dp/0226501175/ref=sr_1_4?s=books&ie=UTF8&qid=1532889253&sr=1-4&keywords=Machiavellian+intelligence. 978-0226501178.

23. *Mistakes when deceiving.* **Ekman, P.** 1, 1981, Annals of the New York Academy of Sciences, Vol. 364, pp. 269-278.

24. *Receiver psychology and the evolution of animal signals.* **Guilford, T. and Dawkins, M. S.** 1, 1991, Animal Behaviour, Vol. 42, pp. 1-14.

25. **Erikson, E. H.** *Childhood and Society.* New York : W. W. Norton & Company, 1993/1950. p. 445. 978-0393310689.

26. —. Growth and Crisis of the Healthy Personality.p 51-107. *Identity and the Life Cycle.* New York : W, Norton nd Company Inc., 1994/1950, Vol. 1, p. 192.

27. **Fitzgerald, H. F., Karraker, K. H. and Luster, T.** *Infant Development: Ecological Perspectives.* New York : Routledge Palmer, 2016. p. 340. 978-1138972704.

28. **Hall, E. T.** *The Silent Language.* New York : Anchor Books, 1959/1973. p. 217. 978-0385055499.

29. **Jackson, P.** *Life in Classrooms.* New York : Teachers College Press, 1968/1990. p. 168. 978-0807730348.

30. *Neurobiology of empathy and callousness: implications for the development of antisocial behavior.* **Shirtcliff, E. A., et al.** 2, 2009, Behavioral Science Law, Vol. 27, pp. 131-171.

31. *They know the words, but not the music: Affective and semantic priming in individuals with psychopathy.* **Blair, K. S., et al.** 2, 2010, Biological Psychology, Vol. 73, pp. 114-123.

32. *Faces and facets: The variability of emotion recognition in psychopathy reflects its affective and antisocial features.* **Igoumenou, A., et al.** 8, 2017, Journal of Abnormal Psychology, Vol. 126, pp. 1066-1076.

33. *Abnormal processing of affective words by psychopaths.* **Williamson, S., Harpur, T. J. and Hare, R . D.** 3, 1991, Psychophysiology, Vol. 28, pp. 260-273.

34. *Corpus Callosum Abnormalities in Psychopathic Antisocial Individuals.* **Raine, A., et al.** 11, 2003, Archives of General Psychiatry, Vol. 60, pp. 1134-1142.

35. *A brain imaging (single photon emission computerized tomography) study of semantic and affective processing in psychopaths.* **Intratorab, J., et al.** 2, 1997, Biological Psychiatry, Vol. 42, pp. 96-103.

36. *The evolution of cooperation.* **Axelrod, R. and Hamilton, W. D.** 4489, 1981, Science, Vol. 211, pp. 1390-1396.

37. *Neural correlates of the judgment of lying: A functional magnetic resonance imaging study.* **Harada, T., et al.** 1, 2009, Neuroscience Research, Vol. 63, pp. 24-34.

38. *Orchid Sexual Deceit Provokes Ejaculation.* **Gaskett, A. C., Winnick, C. G. and Herberstein, M. E.** 6, 2008, American Naturalist, Vol. 171, pp. E206-E212.

39. *Colour mimicry and sexual deception by Tongue orchids (Cryptostylis).* **Gaskett, A. C. and Herberstein, M. E.** 1, 2010, Naturwissenschaften, Vol. 97, pp. 97-102.

40. *Light moulds plastic brains.* **Thor, S.** 13, 2008, Nature, Vol. 456, pp. 177-178.

41. *Molecular Aspects of Monogamy.* **Insel, T. R., Young, L. and Wang, Z.** 1, 1997, Annals of the New York Academy of Sciences, Vol. 807, pp. 302–316.

42. *Oxytocin and the molecular basis of monogamy.* **Insel, T. R., et al.** 1995, Advances in Experimental Medicine and Biology, Vol. 395, pp. 227-234.

43. *Vasopressin-dependent neural circuits underlying pair bonding in the monogamous prairie vole.* **Lim, M. M. and Young, L. J.** 1, 2004, Neuroscience, Vol. 125, pp. 35-45.

44. *Facilitation of affiliation and pair-bond formation by vasopressin receptor gene transfer into the ventral forebrain of a monogamous vole.* **Pitkow, L. J., et al.** 18, 2001, Journal of Neuroscience, Vol. 21, pp. 7392-7396.

45. *Peptides of love and fear: vasopressin and oxytocin modulate the integration of information in the amygdala.* **Dębiec, J.** 9, 2005, BioEssays, Vol. 27, pp. 869-873.

46. *Vasopressin in the lateral septum regulates pair bond formation of male prairie voles.* **Lui, Y., Curtis, J. T. and Wang, Z. X.** 4, 2001, Behavioral Neuroscience, Vol. 115, pp. 910-919.

47. *Roles of Oestrogen Receptors and β in Behavioural Neuroendocrinology: Beyond Yin/Yang.* **Rissman, E. F.** 6, 2008, Journal of Neuroendocrinology, Vol. 20, pp. 873-879.

48. *A gender-specific mechanism for pair bonding: Oxytocin and partner preference formation in monogamous voles.* **Insel, T. R. and Hulihan, T. J.** 4, 1995, Behavioral Neurosciences, Vol. 109, pp. 782-789.

49. *Patterns of brain vasopressin receptor distribution associated with social organization in microtine rodents.* **Insel, T. R., Wang, Z. X. and Ferris, C. F.** 9, 1994, Journal of Neuroscience, Vol. 14, pp. 5381-5392.

50. *Oxytocin, vasopressin, and the neuroendocrine basis of pair bond formation.* **Insel, T. R., et al.** 1998, Advances in Experimental Medicine and Biology, Vol. 449, pp. 215–224.

51. *An evolutionarily conserved form of gonadotropin-releasing hormone coordinates energy and reproductive behavior.* **Temple, J. L., Millar, R. P. and Rissman, E. F.** 1, 2003, Endocrinology, Vol. 144, pp. 13-19.

52. *Effects of gonadotropin-releasing hormones, corticotropin-releasing hormone, and vasopressin on female sexual behavior.* **Schiml, P. A. and Rissman, E. F.** 3, 2000, Hormones and Behavior, Vol. 37, pp. 212-220.

53. *On optimal use of a patchy environment.* **MacArthur, R. H. and Pianka, E. R.** 916, 1966, American Naturalist, Vol. 100, pp. 603-609.

54. *The paradoxical nature of hunter-gatherer diets: Meat-based, yet non-atherogenic.* **Cordain, L., et al.** Suppl 1, 2002, European Journal of Clinical Nutrition, Vol. 56, pp. S42-S52.

55. **Cordain, L., Watkins, B. A. and Mann, N. J.** Fatty acid composition and energy density of foods available to African Hominids. p 144-161. [book auth.] A. P. Simopoulos and K. N. Pavlou. *Nutrition and Fitness: Metabolic Studies in Health and Disease, Athens.* Basel : S. Karger, 2001, Vol. 90, p. 198. World Review of Nutrition and Dietetics.

56. *Overlapping and interactive pathways regulating appetite and craving.* **Kalra, S. P. and Kalra, P. S.** 3, 2004, Journal of Addictive Diseases, Vol. 23, pp. 5-21.

57. *Adipose Tissue Resistin Expression Is Severely Suppressed in Obesity and Stimulated by Peroxisome Proliferator-activated Receptor γ Agonists.* **Way, J. M., et al.** 2001, The Journal of Biological Chemistry, Vol. 276, pp. 25651-25653.

58. *Omental and Subcutaneous Adipose Tissues of Obese Subjects Release Interleukin-6: Depot Difference and Regulation by Glucocorticoid.* **Fried, S. K., Bunkin, D. A. and Greenberg, A. S.** 3, 1998, The Journal of Clinical Endocrinology & Metabolism, Vol. 8, pp. 847-850.

59. *Adipose expression of tumor necrosis factor-alpha: direct role in obesity-linked insulin resistance.* **Hotamisligil, G. S., Shargill, N. S. and Spiegelman, B. M.** 5091, 1993, Science, Vol. 259, pp. 87-91.

60. *Resistin, central obesity, and type 2 diabetes.* **McTernan, C. L., et al.** 9300, 2002, Lancet, Vol. 359, pp. 46–47.

61. *Control of food consumption by learned cues: A forebrain–hypothalamic network.* **Petrovich, G. D. and Gallagher, M.** 4, 2007, Physiology & Behavior, Vol. 91, pp. 397-403.

62. **Dahl, R.** *Charlie and the Chocolate Factory.* reprint edition. New York : Puffin Books, Penguin Random House, 1964/2017. p. 192. 978-0142410318.

63. *Activation of hypothalamic β-endorphin pools by reward induced by highly palatable food.* **Dum, J., Gramsch, C. and Herz, A.** 3, 1983, Pharmacology Biochemistry and Behavior, Vol. 18, pp. 443-447.

64. *Fish schooling: A possible factor in reducing predation.* **Brock, V. E. and Riffenburgh, R. H.** 3, 1960, Journal du Conseil permanent International pour l' Exploration de la Mer, Vol. 25, pp. 307-317.

65. *Some observations of the behavior of schools of fishes in the Gulf of Mexico and adjacent waters.* **Springer, S.** 1, 1957, Ecology, Vol. 38, pp. 166-171.

66. *Insect mimicry of plants dates back to the Permian.* **Garrouste, R., et al.** s.l. : 7, 2016, Nature Communication. Article number: 13735 .

67. *Notes on the biology and ecology of the leaf-mimicking katydid Typophyllum bolivari Vignon (Orthoptera Tettigoniidae: Pseudophyllinae: Pterochrozini).* **Castner, J. L. and Nickle, D. A.** 1995, Journal of Orthoptera Research, Vol. 4, pp. 105-109.

68. *Selection experiments on industrial melanism in the Lepidoptera.* **Kettlewell, H. B. D.** 1955, Heredity, Vol. 9, pp. 321-342.

69. *Further selection experiments on industrial melanism in the Lepidoptera.* **Kettlewell, H. B. D.** 1956, Heredity, Vol. 10, pp. 287-301.

70. *Darwin's Missing Evidence.* **Kettlewell, H. B. D.** 3, 1959, Scientific American, Vol. 200, pp. 48-53.

71. **Kettlewell, H. B. D.** *The Evolution of Melanism: The Study of a Recurring Necessity, with Special Reference to Industrial Melanism in the Lepidoptera.* Oxford : Clarendon Press, 1973. p. 448. 978-0198573708.

72. *Industrial Melanism in British Peppered Moths Has a Singular and Recent Mutational Origin.* **van't Hof, A. E., et al.** 6032, 2011, Science, Vol. 332, pp. 958-960.

73. *Go-slow" Signalling and the Problem of Automimicry.* **Guilford, T.** 3, 1994, Journal of Theoretical Biology, Vol. 170, pp. 311-316.

74. *Host nest site choice depends on risk of cuckoo parasitism in magpie hosts.* **Expósito-Granados, M., et al.** 6, 2017, Behavioral Ecology, Vol. 28, pp. 1492–1497.

75. *Cuckoo adaptations: Trickery and tuning.* **Davies, N. B.** 1, 2011, Journal of Zoology, Vol. 284, pp. 1-14.

76. *Nest prospecting brown-headed cowbirds 'parasitize' social information when the value of personal information is lacking.* **White, D. J., et al.** 1861, 2017, Proceedings of the Royal Society B., Vol. 284.

77. *Retaliatory mafia behaviour by a parasitic cowbird favors host acceptance of parasitic eggs.* **Hoover, J. P. and Robinson, S. K.** 11, 2007, Proceedings of the National Academy of Science, Vol. 104, pp. 4479-4483.

78. **Soler, M., Pérez-Contreras, T. and Soler, J. J.** Brood Parasites as Predators: Farming and Mafia Strategies. [book auth.] M Soler. *Avian Brood Parasitism.* s.l. : Springer, Cham, 2017, pp. 271-286.

79. *Change in host rejection behavior mediated by the predatory behavior of its brood parasite.* **Soler, J. J., et al.** 3, 1999, Behavioral Ecology, Vol. 10, pp. 275-280.

80. *Egg pecking and puncturing behaviors in shiny and screaming cowbirds: Effects of eggshell strength and degree of clutch completion.* **Cossa, N. A., et al.** 60, 2017, Behavioral Ecology and Sociobiology, Vol. 71.

81. *Frequency-dependent Batesian mimicry.* **Pfennig, D. W., Harcombe, W. R. and Pfennig, K. S.** 6826, 2001, Nature, Vol. 41, p. 323.

82. *Contemporary patterns in a historical context: phylogeographic history of the pipevine swallowtail, Battus philenor (papilionidae).* **Fordyce, J. A. and Nice, C. C.** 5, 2003, Evolution, Vol. 57, pp. 1089–1099.

83. *Climate Change and Evolutionary Adaptations at Species' Range Margins.* **Hill, J. K., Griffiths, H. M. and Thomas, C. D.** 2011, Annual Review of Entomology, Vol. 56, pp. 143-159.

84. *Learning of salient prey traits explains Batesian mimicry evolution.* **Kazemi, B., et al.** 3, 2018, Evolution, Vol. 72, pp. 531-539.

85. *Biased generalization of salient traits drives the evolution of warning signals.* **Gamberale-Stille, G., et al.** 1877, 2018, Proceedings of the Royal Society B, Vol. 285.

86. *Conditioned food preferences.* **Sclafani, A.** 2, 1991, Bulletin of the Psychonomic Society, Vol. 29, pp. 256–260.

87. *Why Have Multiple Plastic Responses? Interactions between Color Change and Heat Avoidance Behavior in Battus philenor Larvae.* **Nielson, M. E. and Papaj, D. R.** 6, 2017, The American Naturalist, Vol. 189, pp. 657-666.

88. *Clicking caterpillars: Acoustic aposematism in Antheraea polyphemus and other Bombycoidea.* **Brown, S. G., Boettner, G. H. and Yack, J. E.** 6, 2007, Journal of Experimental Biology, Vol. 210, pp. 993-1005.

89. *Myrmecomorphy: Morphological and behavioral mimicry of ants.* **McIver, J. D. and Stonedahl, G.** 1, 1993, Annual Review of Entomology, Vol. 38, pp. 51-79.

90. *Multi-trait mimicry of ants by a parasitoid wasp.* **Malcicka, M., et al.** 2015, Scientific Reports, Vol. 5, p. 8043.

91. *Ant-Mimicking Spiders: Strategies for Living with Social Insects.* **Ceccarelli, F. S.** 2013, Psyche, p. 839181.

92. *Myrmecomorphy and myrmecophily in spiders: A review.* **Cushing, P. E.** 2, 1997, Florida Entomologist, Vol. 50, pp. 165-193.

93. *Spider-ant associations: an updated review of myrmecomorphy, myrmecophily, and myrmecophagy in spiders.* **Cushing, P. E.** 2012, Psyche: A Journal of Entomology, Vol. 2012, p. 151989.

94. *Aggressive use of Batesian mimicry by an ant-like jumping spider.* **Nelson, X. J. and Jackson, R. R.** 6, 2009, Biological Letters, Vol. 5, pp. 755–757.

95. *Deep-time convergence in rove beetle symbionts of army ants.* **Maruyama, M. and Parker, J.** 6, 2017, Current Biology, Vol. 27, pp. P920-926.

96. *The role of ultraviolet colour in the assessment of mimetic accuracy between Batesian mimics and their models: A case study using ant-mimicking spiders.* **Corcobado, G., Herberstein, M. E. and Pekár, S.** 2016, The Science of Nature, Vol. 103, p. 90.

97. *Queen ants make distinctive sounds that are mimicked by a butterfly social parasite.* **Barbero, F., et al.** 5915, 2009, Science, Vol. 373, pp. 782–785.

98. *The Pied Piper: A parasitic beetle's melodies modulate ant behaviours.* **Di Guilio, A., et al.** 7, 2015, PLOS-1, Vol. 10, p. e0130541.

99. *Arthropods Associate with their Red Wood ant Host without Matching Nestmate Recognition Cues.* **Parmentier, T., Dekoninck, W. and Wenseleers, T.** 7, 2017, Journal of Chemical Ecology, Vol. 43, pp. 644-661.

100. *Seed dispersal by ants: Behaviour-releasing compounds in elaiosomes.* **Brew, C. R., O'Dowd, D. J. and Rae, I. D.** 4, 1989, Oecologia, Vol. 80, pp. 490-497.

101. *Myrmecophilous aphids produce cuticular hydrocarbons that resemble those of their tending ants.* **Endo, S. and Itino, T.** 1, 2013, Population Ecology, Vol. 55, pp. 27-34.

102. *Aggressive mimicry coexists with mutualism in an aphid.* **Salazar, A., et al.** 4, 2015, Proceedings of the National Academy of Science of the United States of America, Vol. 112, pp. 1101-1106.

103. *The adaptive bases of ant-mimicry in a neotropical aphantochilid spider (Araneae: Aphantochilidae).* **Olivera, P. S. and Sazima, I.** 2, 1984, Biological Journal of the Linnean Society, Vol. 22, pp. 145–155.

104. *Ant mimicry in the spider Myrmecotypus iguazu (Araneae: Corinnidae), with notes about myrmecomorphy in spiders.* **Rubio, G. D., Arbino, M. O. and Cushing, P. E.** 3, 2013, The Journal of Arachnology, Vol. 41, pp. 395-399.

105. *Walking like an ant: A quantitative and experimental approach to understanding locomotor mimicry in the jumping spider Myrmarachne formicaria.* **Shamble, P. S., et al.** 1858, 2017, Proceedings of the Royal Society B, Vol. 284.

106. *A cuckoo-like parasitic moth leads African weaver ant colonies to their ruin.* **Dejean, A., et al.** 2016, Scientific Reports, Vol. 6, p. 23778.

107. *Higher productivity at the cost of increased host-specificity when Maculinea butterfly larvae exploit ant colonies through trophallaxis rather than by predation.* **Thomas, J. A. and Elmes, G. W.** 4, 2004, Ecological Entomology, Vol. 23.

108. *The "Evolutionarily Significant Unit" concept and its applicability in biological conservation.* **Casacci, L. P., Barbero, F. and Balletto, E.** 2, 2014, Italian Journal of Zoology, Vol. 81, pp. 182-193.

109. *A Batesian ant-mimicry complex from the Mountain Pine Ridge of British Honduras, with an example of transformational mimicry.* **Jackson, J. F. and Drummond, B. A.** 1, 1974, American Midland Naturalist, Vol. 91, pp. 248-251.

110. *Insects as flower visitors and pollinators.* **Kevan, P. G. and Baker, H. G.** 1, 1983, Annual Review of Entomology, Vol. 28, pp. 407-453.

111. *Microdon larvae in Pseudomyrma nests.* **Wheeler, W. M.** 303, 1901, Psyche, Vol. 9, pp. 223-224.

112. *Social encapsulation of beetle parasites by Cape honeybee colonies (Apis mellifera capensis Esch.).* **Neumann, P., et al.** 5, 2001, Naturwissenschaften, Vol. 88, pp. 214-216.

113. *Small hive beetles survive in honeybee prisons by behavioural mimicry.* **Ellis, J. D., et al.** 7, 2002, Naturwissenschaften, Vol. 89, pp. 326-328.

114. **Carrol, L. and Gardner, M.** *The Annotated Alice: The Definitive Edition: Through the Looking Glass and What Alice Found There.* New York : W. Norton & Company, 1960/2000. p. 352.

115. *1.* **Van Valen, L.** 1973, Evolutionary Theory, pp. 1-30.

116. *The complex business of survival by aposematism.* **Mappes, J., Marples, N. and Endler, J. A.** 11, 2005, Trends in Ecology & Evolution, Vol. 20, pp. 598-603.

117. *Learning about aposematic prey.* **Skelhorn, J., Halpin, C. G. and Rowe, C.** 4, 2016, Behavioral Ecology, Vol. 27, pp. 955–964.

118. *Why are warning displays multimodal?* **Rowe, C. and Halpin, C.** 9, 2013, Behavioral Ecology and Sociobiology, Vol. 67, pp. 1425-1439.

119. *Crying wolf to a predator: Deceptive vocal mimicry by a bird protecting young.* **Igic, B., et al.** 1809, 2015, Proceedings of the Royal Society B. Biological Sciences, Vol. 282, p. 20150798.

120. *Biomechanics of a moth scale at ultrasonic frequencies.* **Shen, Z., et al.** 48, 2018, Proceedings of the National Academy of Science, Vol. 115, pp. 12200-12205.

121. *Acoustic mimicry in a predator–prey interaction.* **Barber, J. R. and Conner, W. E.** 22, 2007, Proceedingsof the National Academy of Sciences of the United States of America, Vol. 104, pp. 9331-9334.

122. *Biased generalization of salient traits drives the evolution of warning signals.* **Gamberale-Stille, G., et al.** 1877, 2018, PSroceedings of the Royal Society B, Vol. 285, p. 20180283.

123. *Mantids and milkweed bugs: Efficacy of aposematic coloration against invertebrate predators.* **Berenbaum, M. R. and Miliczky, E.** 1, 1984, American Midland Naturalist, Vol. 111, pp. 64-68.

124. *Birds learn to use distastefulness as a signal of toxicity.* **Skelhorn, J. and Rowe, C.** 1688, 2010, Proceedings of the Royal Society B, Vol. 277, pp. 1729-1734.

125. *Nutrient sensing, taste and feed intake in avian species.* **Niknafs, S. and Roura, E.** 2018, Nutrition Research Reviews, pp. 1-11.

126. *An individual-based profitability spectrum for understanding interactions between predators and their prey.* **Marples, N. M., Speed, M. P. and Thomas, R. J.** 2018, Biological Jorunal of the Linnean Society, p. bly088.

127. *High-model abundance may permit the gradual evolution of Batesian mimicry: an experimental test.* **Kikuchi, D. W. and Pfennig, D. W.** 1684, 2010, Proceedings of the Royal Society of London B: Biological Sciences, Vol. 277, pp. 1041-1048.

128. *Coral snakes predict the evolution of mimicry across New World snakes.* **Rabosky, A. R. D., et al.** 2016, Nature Communications, Vol. 7, p. 11484.

129. *Frequency dependence shapes the adaptive landscape of imperfect Batesian mimicry.* **Finkbinder, S. D., et al.** 1876, 2018, Proceedings of the Royal Society B: Biological Sciences, Vol. 285, p. 20172786.

130. *Dense white trichome production by plants as possible mimicry of arthropod silk or fungal hyphae that deter herbivory.* **Yamazaki, K. and Lev-Yadun, S.** 7, 2015, ScienceDirect, Vol. 364, pp. 1-6.

131. *Darwin, Wallace, and the origin of sexual dimorphism.* **Kottler, M. J.** 3, 1980, Proceedings of the American Philosophical Society, Vol. 124, pp. 203-226.

132. *Why are female color polymorphisms rare in territorial damselflies?* **Castillo, M. B. C. and Grether, G. F.** 9, 2018, Ethology, Vol. 124, pp. 667–673.

133. *A role for ecology in the evolution of colour variation and sexual dimorphism in Hawaiian damselflies.* **Cooper, I. A, Brown, J. M. and Getty, T.** 2, 2018, Journal of Evolutionary Biology, Vol. 29, pp. 418-427.

134. *Mimetic butterflies support Wallace's model of sexual dimorphism.* **Kunte, K.** 1643, 2008, Proceedings of the Royal Society of London B: Biological Sciences, Vol. 275, pp. 1617-1624.

135. *Chapter 3 Invasion of the Body Snatchers: The Diversity and Evolution of Manipulative Strategies in Host–Parasite Interactions.* **Lefèvre, Thierry, et al.** 2009, Advances in Parasitology, Vol. 68, pp. 45-83.

136. *Parasitism, personality and cognition in fish.* **Barber, I., et al.** 3, 2017, Behavioural Processes, Vol. 141, pp. 205-219.

137. *Modification of host social networks by manipulative parasites.* **Poulin, R.** s.l. : Brill, 2017, Behaviour, Vol. 154.

138. *Two steps to suicide in crickets harbouring hairworms.* **Sanchez, M. I., et al.** 5, 2008, Animal Behaviour, Vol. 76, pp. 1621-1624.

139. *Sex-dependent neurotransmitter level changes in brains of Toxoplasma gondii infected mice.* **Gatkowska, J., et al.** 1, 2013, Experimental Parasitology, Vol. 133, pp. 1-7.

140. *Parasite-infected sticklebacks increase the risk-taking behaviour of uninfected group members.* **Demandt, N,, et al.** 1881, 2018, Proceedings of the Royal Academy B: Biological Sciences, Vol. 285, p. 20180956.

141. *Behaviour of roach (Rutilus rutilus L.) altered by Ligula intestinalis (Cestoda: Pseudophyllidea): a field demonstration.* **Loot, G., et al.** 9, 2008, Freshwater Biology Volume, Vol. 46, pp. 1219–1227.

142. *The differential effects of Ligula intestinalis (L.) plerocercoids on host growth in three natural populations of roach, Rutilus rutilus (L.).* **Loot, G., et al.** 3, 2002, Ecology of Freshwater Fish, Vol. 11, pp. 168–177.

143. *Malaria infection and human evolution.* **Sabbatani, S., Manfredi, R. and Fiorino, S.** 1, 2010, Le Infezioni in Medicina : Rivista Periodica di Eziologia, Epidemiologia, Diagnostica, Clinica e Terapia delle Patologie Infettive, Vol. 18, pp. 56-74. PMID: 20424529.

144. **Department of Information, Evidence and Research, WHO.** *Causes of Death.* Geneva : World Health Organization, 2004.

145. —. *WHO Methods and Data Sources for Country-Level Causes of Death 2000-2015.* Geneva : World Health Organization, 2017.

146. *Waiting with bated breath: opportunistic orientation to human odor in the malaria mosquito, Anopheles gambiae, is modulated by minute changes in carbon dioxide concentration.* **Webster, B., Lacey, E. S. and Cardé, R. T.** 1, 2015, Journal of Chemical Ecology, Vol. 41, pp. 59-66.

147. *Realistic imitation of mosquito's proboscis: Electrochemically etched sharp and jagged needles and the cooperative inserting motion.* **Izumi, H., et al.** 2011, 2009, Sensors and Actuators: Special Issues: Transducers 2009, Vol. A165, pp. 115-123.

148. *Mosquito proboscis: An elegant biomicroelectromechanical system.* **Kong, X. Q. and Wu, C. W.** 1, 2010, Physical Review, E, Vol. 82, p. 011910.

149. *Nocturnal Itch: Why Do We Itch At Night?* **Patel, T., Ishiuji, Y. and Yosipovitch, G.** 2007, Acta Dermato-Venereologica, Vol. 87, pp. 295–298.

150. *Time-dependent variations of the skin barrier function in humans: transepidermal water loss, stratum corneum hydration, skin surface pH, and skin temperature.* **Yosipovitch, G., et al.** 1, 1998, The Journal of Investigative Dermatology, Vol. 110, pp. 20–23.

151. **Ewing, A. W.** *Anthropod Bioacouctics.* Edinburgh : Edinburgh University Press, 1989.

152. **Gelfand, S. A.** *Hearing: An Introduction to Psychological and Physiological Acoustics.* Boca Raton : CRC Press, Taylor and Francis Group, 2017. p. 417. 978-1498775427.

153. *Tertian and Quartan Fevers: Temporal Regulation in Malarial Infection.* **Garcia, C. R. S., Markus, R. P. and Madeira, L.** 5, 2001, Journal of Biological Rhythms, Vol. 16, pp. 436-443.

154. *Calcium-dependent modulation by melatonin of the circadian rhythm in malarial parasites.* **Hotta, C. T., et al.** 1, 2000, Nature Cell Biology, Vol. 2, pp. 466-468.

155. *Melatonin and N-acetyl-serotonin cross the red blood cell membrane and evoke calcium mobilization in malarial parasites.* **Hotta, C. T., Markus, R. P. and Garcia, C. R. S.** 11, 2003, Brazilian Journal of Medical and Biological Research, Vol. 36, pp. 1583-1587.

156. *Daily rhythms in mosquitoes and their consequences for malaria transmission.* **Rund, S. S. C., et al.** 2, 2016, Insects, Vol. 7, p. 14.

157. *Competition for red blood cells can enhance Plasmodium vivax parasitemia in mixed-species malaria infections.* **McQueen, P. G. and McKenzie, F. E.** 1, 2006, The American Journal of Tropical Medicine and Hygiene, Vol. 75, pp. 112–125.

158. *Competition and evolution of reproductive restraint in malaria parasites.* **Pollitt, L. C., et al.** 3, 2011, American Naturalist, Vol. 177, pp. 358-367.

159. **Singer, B.** *House: Pilot Episode.* [writ.] D. Shore. Heel and Toe Films; Shore Z Productions; Bad Hat Harry Productions, NBC Universial Productiosn; Fox Network; CanWest Global Television, 2004.

160. *An extraordinary reproductive strategy in freshwater bivalves: prey mimicry to facilitate larval dispersal.* **Haag, W. R., Butler, R. S. and Hartfield, P. D.** 3, 1995, Freshwater Biology, Vol. 34, pp. 471-476.

161. **O'Brian, C.** *Monograph (1287): Reproductive Biology of Amblema neislerii, Elliptoideus sloatianus, Lampsilis subangulata, Medionidus penicillatus, and Pleurobema pyriforme (Bivalvia: Unionidae).* Gainesville : Florida Caribbean Science Center, U. S. Geological Survey, 1997. p. 70pp.

162. *Possible Mimicry in the Galeommatid Bivalve Scintilla philippinensis (Bivalvia: Galeommatidae) to Nudibranchs and Crabs.* **Ohtsuka, S, et al.** 1-4, 2017, Venus (Journal of the Malacological Society of Japan), Vol. 75, pp. 93-98.

163. *Host attraction, brooding phenology, and host specialization on freshwater drum by 4 freshwater mussel species.* **Sietman, B. E., Hove, M.C. and Davis, J. M.** 1, 2018, Freshwater Science, Vol. 37, pp. 96-107.

164. *A protective function for aggressive mimicry?* **Côté, I. M. and Cheney, K. L.** 1624, 2007, Proceedings of the Royal Society B, Vol. 274, pp. 2445–2448.

165. *Mimicry in Tropical Fishes.* **Wickler, W.** 772, 1966, Philosophical Transactions of the Royal Society of London. Series B, Biological Sciences, Vol. 251, pp. 473-474.

166. *The evolution of fangs, benom, and mimicry systems in blenny fishes.* **Casewell, N. R., et al.** 8, 2017, Current Biology, Vol. 27, pp. 1184-1191.

167. *To females of a noctuid moth, male courtship songs are nothing more than bat echolocation calls.* **Nakano, R., et al.** 5, 2010, Biological Letters, Vol. 6, pp. 582-584.

168. *Evolution of deceptive and true courtship songs in moths.* **Nakano, R., et al.** 2013, Scientific Reports, Vol. 3, p. 2003 .

169. *Behavioral Mimicry in the Titmice (Paridae) and Certain Other Birds.* **Sibley, C. G.** 2, 1955, The Wilson Bulletin, Vol. 67, pp. 128-132.

170. *Mimicry of friarbirds by oriels.* **Diamond, J. M. M.** 2, 1982, The Auk, Vol. 99, pp. 187-196.

171. *Repertoire Size and Syllable Sharing in the Song of the Clay-Colored Thrush (Turdus grayi).* **Vargas-Castro, L. E., Sánchez, N. V. and Barrantes, G.** 3, 2012, The Wilson Journal of Ornithology, Vol. 124, pp. 446-453.

172. *The vocal behavior of the Indian hill mynah, Gracula religiosa.* **Bertram, B.** 2, 1970, Animal Behavior Monograph, Vol. 3, pp. 79-192.

173. *Wild mynahs mimic wild primates.* **Tenaza, R. R.** 5544, 1976, Nature, Vol. 259, p. 561.

174. **Byrne, R. W. and Whiten, A.** Computation and mindreading in primate tactical deception. [book auth.] A. Whiten. *Natural Theories of Mind: Evolution, Development and Simulation of Everyday Mindreading.* s.l. : Blackwell Publishing, 1991, p. 384.

175. *Cognitive evolution in primates: evidence from tactical deception.* **Byrne, R. W. and Whiten, A.** 3, 1992, Man, Vol. 27, pp. 609-627.

176. *Do chimpanzees know what conspecifics know?* **Hare, B., Call, J. and Tomasello, M.** 1, 2001, Animal Behaviour, Vol. 61, pp. 139-151.

177. *Does the chimpanzee have a theory of mind?* **Premack, D. and Woodruff, G.** 4, 1978, Behavioral and Brain Sciences, Vol. 1, pp. 515-526.

178. *The roots of empathy: Through the lens of rodent models.* **Meyza, K. Z., et al.** Part B, 2017, Neuroscience & Biobehavioral Reviews, Vol. 76, pp. 216-234.

179. **Godfrey-Smith, P.** *Other Minds: The Octopus, the Sea, and the Deep Origins of Consciousness.* New York : Farrar, Straus and Giroux, 2016. 978-0374227760.

180. *Octopuses (Enteroctopus dofleini) recognize individual humans.* **Anderson, R. C., et al.** 3, 2010, Journal of Applied Animal Welfare Science, Vol. 13, pp. 261-272.

181. *What insects can tell us about the origins of consciousness.* **Barron, A. B. and Klein, C.** 2016, Proceedings of the National Academy of Science of the United States of America, p. 20152008.

182. *Insect consciousness: Fine-tuning the hypothesis.* **Mallatt, J. and Feinberg, T. E.** 2016, Animal Sentience, Vol. 2016.

183. *Information transfer beyond the waggle dance: Observational learning in bees and flies.* **Avarguès-Weber, A., et al.** 2015, Frontiers in Ecology and Evolution, Vol. 12.

184. *Bumblebees show cognitive flexibility by improving on an observed complex behavior.* **Loukola, O. J., et al.** 6327, 2017, Science, Vol. 355, pp. 833-836.

185. *Detecting cheaters without thinking: Testing the automaticity of the cheater detection module.* **Van Lier, J., Revlin, R. and Neys, W. D.** 1, 2013, PloS One, Vol. 8, p. e53827.

186. *Behavioral spillovers from freeriding in multilevel interactions.* **Thommes, K., Vyrastekova, J. and Akkerman, A.** 1, 2015, Journal of Behavioral and Experimental Economics, Vol. 56, pp. 78-87.

187. *How chimpanzees cooperate in a competitive world.* **Suchak, M., et al.** 36, 2016, Proceedings of the National Academy of Sciences, Vol. 113, pp. 10215-10220.

188. *The Chimpanzee's service economy: Food for grooming.* **de Waal, F. B. M.** 6, 1997, Evolution and Human Behavior, Vol. 18, pp. 375-386.

189. *Do tufted capuchin monkeys (Cebus apella) spontaneously deceive opponents? A preliminary analysis of an experimental food-competition contest between monkeys.* **Fujita, K., Kuroshima, H. and Masuda, T.** 1, 2002, Animal Cognition, Vol. 65, pp. 19-25. PMID:.

190. *Female monkeys use both the carrot and the stick to promote male participation in intergroup fights.* **Arseneau-Robar, T. J. M., et al.** 1843, 2016, Proceedings of the Royal Society B: Biological Sciences, Vol. 283, p. Article ID:20161817.

191. *Real-time social selection maintains honesty of a dynamic visual signal in cichlid fish.* **Bachmann, J. C., et al.** 5, 2017, Evolution Letters, Vol. 1, pp. 269-278.

192. *Generalized rule application in bluestreak cleaner wrasse (Labroides dimidiatus): Using predator species as social tools to reduce punishment.* **Wismer, S., Grutter, A. and Bshary, R.** 4, 2016, Animal Cognition, Vol. 19, pp. 769-778.

193. *Can animals recall the past and plan for the future?* **Clayton, N. S., Bussey, T. J. and Dickinson, A.** 8, 2003, Nature Reviews Neuroscience, Vol. 4, pp. 685–691.

194. *Episodic-like memory during cache recovery by scrub jays.* **Clayton, N. S. and Dickinson, A.** 1998, Nature, Vol. volume 395, pp. pages 272–274.

195. *Effects of experience and social context on prospective caching strategies by scrub jays.* **Emery, N. J. and Clayton, N. S.** 22, 2001, Nature, Vol. 414, pp. 443–446.

196. **Emery, N. J. and Clayton, N. S.** How to Build a Scrub-Jay that Reads Minds. [ed.] S. Itakura and K. Fujita. *Origins of the Social Mind Part 1.* Tokyo : s.n., 2008, pp. 65-97.

197. *Western scrub-jays (Aphelocoma californica) use cognitive strategies to protect their caches from thieving conspecifics.* **Emery, N. J., Dally, J. M. and Clayton, N. S.** 1, 2004, Animal Cognition, Vol. 7, pp. 37-43.

198. *Problems faced by food-caching corvids and the evolution of cognitive solutions.* **Grodzinski, U. and Clayton, N. S.** 1542, 2010, Philosophical Transactions of the Royal Society of London B: Biological Sciences, Vol. 365, pp. 977-987.

199. *Ravens judge competitors through experience with play caching.* **Bugnyar, T., et al.** 20, 2007, Current Biology, Vol. 17, pp. 1804-1808.

200. *Ravens, Corvus corax, differentiate between knowledgeable and ignorant competitors.* **Bugnyar, T. and Heinrich, B.** 1573, 2005, Proceedings of the Royal Society B, Vol. 272, pp. 1641–1646.

201. *Pilfering ravens, Corvus corax, adjust their behaviour to social context and identity of competitors.* **Bugnyar, T. and Heinrich, B.** 4, 2008, Animal Cognition, Vol. 9, pp. 369-376.

202. *Effects of Minor Food Theft on Social Perceptions of Culprits and Victims.* **Alley, T. R.** 1, 2018, Evolutionary Psychological Science, Vol. 4, pp. 90–97.

203. **Alley, T. R. and Kooi, K. A.** Social perception of culprits and victims by witnesses of minor food theft. p 97-100. [ed.] J. A. Weast-Knapp, M. Malone and D. H. Abney. *Studies in Perception and Action XIII: Eighteenth International Conference on Perception and Action.* New York : Psychological Press, 2015, Chapter 3, p. 276.

204. *Rich pickings near large communal roosts favor 'gang' foraging by juvenile common ravens, Corvus corax.* **Dall, S. R. and Wright, J.** 2, 3009, PLoS One, Vol. 4, p. e4530.

205. *Leading a conspecific away from food in ravens (Corvus corax)?* **Bugnyar, T. and Kotrschal, K.** 2, 2004, Animal Cognition, Vol. 7, pp. 69–76.

206. *Audience effects on food caching in grey squirrels (Sciurus carolinensis): Evidence for pilferage avoidance strategies.* **Leaver, L. A., et al.** 1, 2009, Animal Cognition, Vol. 10, pp. 23-27.

207. *Eavesdropping squirrels reduce their future value of food under the perceived presence of cache robbers.* **Schmidt, K. A. and Ostfeld, R. S.** 3, 2008, The American Naturalist, Vol. 171, pp. 386-393.

208. *Cache protection strategies of a scatter-hoarding rodent: do tree squirrels engage in behavioural deception?* **Steele, M. A., et al.** 2, 2008, Animal Behaviour, Vol. 75, pp. 705-714.

209. *Cognition in the African Grey parrot: Preliminary evidence for auditory/focal comprehension of class concept.* **Pepperberg, I. M.** 2, 1983, Ethology, Vol. 11, pp. 179-185.

210. *Phenotypic correlates of scrounging behavior in zebra finches: Role of foraging efficiency an dominance.* **Beauchamp, G.** 9, 2006, Ethology, Vol. 112, pp. 873-878.

211. **Adenzato, M. and Ardito, R. B.** The Role of Theory of Mind and Deontic Reasoning in the Evolution of Deception. pp 7-12. [ed.] M. Hahn and S. C. Stoness. *Proceedings of the Twenty-First Conference of the Cognitive Science Society.* Mahwah : Lawrence Erlbaum Associates, 1999, p. 848. https://www.routledge.com/Proceedings-of-the-Twenty-first-Annual-Conference-of-the-Cognitive-Science/Hahn-Stoness/p/book/9780805835816.

212. *Neocortex size predicts deception rate in primates.* **Byrne, R. W. and Corp, N.** 1549, 2004, Proceedings of the Royal Society. B, Vol. 271, pp. 1693–1699.

213. *Vervet monkey alarm calls: Manipulation through shared information?* **Cheney, D. L. and Seyfarth, R. M.** 1-2, 1985, Behaviour, Vol. 94, pp. 150-166.

214. *The social brain hypothesis and its implications for social evolution.* **Dunbar, R. I.** 5, 2009, Annals of Human Biology, Vol. 36, pp. 562-572.

215. *Understanding primate brain evolution.* **Dunbar, R. I. and Shultz, S.** 1480, 2007, Philosophical Transactions of the Royal Society of London B, Vol. 362, pp. 649-658.

216. *Cues of ownership in longtailed macaques, Macaca fascicularis.* **Kummer, H. and Cords, M.** 4, 1991, Animal Behavior, Vol. 44, pp. 529–549.

217. **Byrne, R. W.** What's the use of anecdotes? Distinguishing psychological mechanisms in primate tactical deception. pp. 134-150. [ed.] R. W. Mitchell, N. S.

Thompson and L. Miles. *Anthropomorphism, Anecdotes, and Animals: The Emperor's New Clothes?* New York : SUNY Press Biology and Philosophy, 1997, p. 538.

218. *Rhesus monkeys, Macaca mulatta, know what others can and cannot hear.* **Santos, L. R., Nissen, A. G. and Ferrugia, J. A.** 1175, 2006, Animal Behaviour, Vol. 71, pp. 1175-1181.

219. *Intentional deception in primates.* **de Waal, F. B. M.** 3, 1992, Evolutionary Anthropology: Issues, News, and Reviews, Vol. 1, pp. 86–92.

220. **de Waal, F. B. M.** *Good Natured: The Origins of Right and Wrong in Humans and Other Animals.* Cambridge : Harvard University Press, 1996. p. 368 . Vol. 1. 978-0674356610.

221. **Menzel, E. W.** A group of young chimpanzees in a 1-acre field: Leadership and communication. pp 83-153. [ed.] A. M. Schrier, H. F. Harlow and F. Stollnitz. *Behavior of Nonhuman Primates: Modern Research Trends.* New York : New York Academic Press, 1974, Vol. 5.

222. *Tactics to obtain a hidden food item in chimpanzee pairs.* **Hirata, S. and Matsuzawa, T.** 3-4, 2001, Animal Cognition, Vol. 4, pp. 285–295.

223. *Variation in withholding of information in three monkey species.* **Amici, F., Call, J. and Aureli, F.** 1671, 2009, Proceedings of the Royal Society B, Vol. 276, pp. 3311-3318. http://dx.doi.org 10.1098/rspb.2009.0759.

224. *Withholding information in semifree-ranging Tonkean macaques (Macaca tonkeana).* **Ducoing, A. M. and Thierry, B.** 1, 2003, Journal of Comparative Psychology, Vol. 117, pp. 67-75.

225. *Learning to outwit a competitor in mangabeys (Cercocebus torquatus torquatus).* **Coussi-Korbel, S.** 2, 1994, Journal of Comparative Psychology, Vol. 108, pp. 164-171.

226. *Sneaky monkeys: an audience effect of male rhesus macaques (Macaca mulatta) on sexual behavior.* **Overduin-Devries, A. M., et al.** 3, 2012, American Journal of Primatology, Vol. 74, pp. 217-228.

227. *Chimpanzees deceive a human competitor by hiding.* **Hare, B., Call, J. and Tomasello, M.** 3, 2006, Cognition, Vol. 101, pp. 495-514.

228. *Spontaneous innovation for future deception in a male chimpanzee.* **Osvath, M. and Karvonen, E.** 5, 2012, PLOS One, Vol. 7, p. 36782.

229. *Chimpanzee uses manipulative gaze cues to conceal and reveal information to foraging competitor.* **Hall, K., et al.** 3, 2017, American Journal of Primatology, Vol. 79, pp. 1-11.

230. *Pointing, withholding information, and deception in capuchin monkeys (Cebus apella).* **Mitchell, R. W. and Anderson, J. R.** 4, 1997, Journal of Comparative Psychology, Vol. 111, pp. 351-361.

231. *Training squirrel monkeys (Saimiri sciureus) to deceive: acquisition and analysis of behavior toward cooperative and competitive trainers.* **Anderson, J. R., et al.** 3, 2001, Journal of Comparative Psychology, Vol. 115, pp. 282–293.

232. *Intentional communication in the chimpanzee: the development of deception.* **Woodruff, G. and Premack, D.** 4, 1979, Cognition, Vol. 7, pp. 333–362.

233. *Monkeys crying wolf? Tufted capuchin monkeys use anti-predator calls to usurp resources from conspecifics.* **Wheeler, B. C.** 1669, 2009, Proceedings of the Royal Society B, Vol. 276, pp. 3013-3018.

234. **Evans, C.** Referential signals. pp. 99–143. [ed.] D. H. Owens, M. D. Beecher and N. S Thompson. *Perspectives in Ethology: Communication* . 1997, 4, p. 431.

235. *Tactical deception in primates.* **Whiten, A. and Byrne, R.** 2, 1988, Behavioral and Brain Sciences, Vol. 11, pp. 233-273.

236. **Byrne, R. W. and Stokes, E. J.** Can monkeys malinger? pp. 52-65. *Malingering & Illness Deception.* Cambridge : Oxford University Press, 2003, p. 384.

237. **Cheney, D. L. and Seyfarth, R. M.** *How Monkeys See the World: Inside the Mind of Another Species.* Chicago, IL : University of Chicago Press, 1990. p. 388. 978-0226102467.

238. **Weber, M.** *On Law in Economy and Society.* New York : Touchstone, 1954/ 1967. p. 363. 978-0671201609.

239. **de Waal, F.B.** *Chimpanzee Politics: Power and Sex Among the Apes.* London : Jonathon Cape, 1982/2007. p. 256. 978-0801886560.

240. *The integration of dominance and social bonding in primates.* **de Waal, F. B. M.** 4, 1986, Quarterly Review of Biology, Vol. 61, pp. 459-479.

241. **Goodall, J.** Unusual violence in the overthrow of an alpha male chimpanzee at Gombe. pp. 131-142. [ed.] T. Nishida, et al. *Topics in Primatology.* Tokyo : University of Tokyo Press, 1992, Vol. 1.

242. *A within-group gang attack on a young adult male chimp: Ostracism of an ill-mannered member.* **Nishida, T., et al.** 2, 1995, Primates, Vol. 36, pp. 207-211.

243. *The fate of defeated alpha male chimpanzees in relation to their social networks.* **Uehara, S., et al.** 1, 1994, Primates, Vol. 35, pp. 49-55.

244. *Neural correlates of the judgment of lying: A functional magnetic resonance imaging study.* **Haradaa, T., et al.** 1, 2009, Neuroscience Research, Vol. 63, pp. 24-34.

245. *The corruption of honest signalling.* **Dawkins, M. S. and Guilford, T.** 1, 1991, Animal Behaviour, Vol. 41, pp. 865-873.

246. *Skeptical responding in Rhesus monkeys (Macaca mulatta).* **Gouzoules, H., Gouzoules, S. and Miller, K.** 4, 1996, International Journal of Primatology, Vol. 17, pp. 549-568.

247. *Cheater detection is modified by social rank: The impact of domianance on the evolution of cognitive function.* **Cummins, D. D.** 4, 1999, Evolution and Human Behaviour, Vol. 20, pp. 229-248.

248. *The selfish nature of generosity: harassment and food sharing in primates.* **Stevens, J. R.** 1538, 2004, Proceedings of the Royal Society, B, Vol. 271, pp. 451-456.

249. *The carrot or the stick: Rewards, punishments and cooperation.* **Andreoni, J. A., Harbaugh, W. and Vesterlund, L.** 3, 2003, American Economic Review,, Vol. 93, pp. 893–902.

250. *Punishment in animal societies.* **Clutton-Brock, T. H. and Parker, G. A.** 1995, Nature, Vol. 373, pp. 209–216.

251. *Evidence for sperm transfer during sneaky mating in wild Trinidadian guppies.* **Matthews, M. and Magurran, A. E.** 6, 2000, Journal of Fish Biology, Vol. 56, pp. 1381-1386.

252. **Murdock, G. P.** *Ethnographic Atlas.* Pittsburgh : University of Pittsburgh Press, 1967. p. 128. ASIN: B0006BR29K.

253. *Polygyny and inheritance of wealth.* **Hartung, J., et al.** 1, 1982, Current Anthropology, Vol. 23, pp. 1-12.

254. *Explaining monogamy and polygyny in human societies: comment on Kanazawa and Still.* **Sanderson, S. K.** 1, 2001, Social Forces, Vol. 80, pp. 329-335.

255. **Malinowski, B.** *The Sexual Life of Savages In North-Western Melanesia; An Ethnographic Account of Courtship, Marriage and Family Life Among the Natives of the Trobriand Islands, British New Guinea.* s.l. : Martino Fine Books, 1929/2012. p. 736. 978-1614272427.

256. *Dowry as female competition.* **Gaulin, S. J. C. and Boster, J. S.** 4, 1990, American Anthropologist, Vol. 92, pp. 994-1005.

257. *Energetic Constraints and Male Mate-Securing Tactics in the Damselfly Calopteryx splendens xanthostoma (Charpentier).* **Plaistow, S. and Siva-Jothy, M. T.** 1374, 1996, Proceedings: Biological Sciences, Vol. 263, pp. 1233-1239.

258. *Territorial defence in a gall aphid.* **Whitham, T. G.** 1979, Nature, Vol. 279, pp. 324–325.

259. *The theory of habitat selection examined and extended using Pemphigus aphids.* **Whitham, T. G.** 4, 1980, American Naturalist, Vol. 115, pp. 449-446.

260. *Costs and benefits of female mate choice: is there a lek paradox?* **Reynolds, J. D. and Gross, M. R.** 2, 1990, American Naturalist, Vol. 136, pp. 230-243.

261. *A possible non-sexual origin of mate preference: are male guppies mimicking food?* **Rodd, F. H., et al.** 1490, 2002, Proceedings of the Royal Society of London, B, Vol. 269, pp. 475-481.

262. *A Cost of Sexual Harassment in the Guppy, Poecilia reticulata.* **Magurran, A. E. and Seghers, B. H.** 1351, 1994, Proceedings of the Royal Society B, Vol. 258, pp. 89-92.

263. *Sex-biased movement in the guppy (Poecilia reticulata).* **Croft, D. P., et al.** 1, 2003, Oecologia, Vol. 137, pp. 62-68.

264. *Male sexually coercive behaviour drives increased swimming efficiency in female guppies.* **Killen, S. S., et al.** 4, 2016, Functional Ecology, Vol. 30, pp. 576-583.

265. *Intra-sexual selection in Drosophila.* **Bateman, A. J.** 3, 1948, Heredity, Vol. 2, pp. 349-368.

266. **Sprengel, C.** *Die Lehre vom Dünger oder Beschreibung aller bei der Landwirthschaft gebräuchlicher vegetablilischer, animalischer und mineralischer Düngermaterialien, nebst Erklärung ihrer Wirkungsart.* Leipzig : s.n., 1839.

267. *Extra-pair mating and evolution of cooperative neighbourhoods.* **Eliassen, S. and Jørgensen, C.** 7, 2014, PLOS One, Vol. 9, p. e99878.

268. **Lewis, E. A.** *Honours Thesis: Cooperative nest defense by European starlings (Sturnus vulgaris) during a predatory threat.* Halifax : Saint Mary's University, 2016. p. 28.

269. *Improving sneaky-sex in a low oxygen environment: Reproductive and physiological responses of male mosquito fish to chronic hypoxia.* **Carter, A. J. and Wilson, R. S.** 24, 2006, Journal of Experimental Biology, Vol. 209, pp. 4878-4884.

270. *Post-mating clutch piracy in an amphibian.* **Vieites, D. R., et al.** 7006, 2004, Nature, Vol. 431, pp. 305-308.

https://scholar.google.com/citations?user=jgWZfZ0AAAAJ&hl=en#d=gs_md_cita-

d&p=&u=%2Fcitations%3Fview_op%3Dview_citation%26hl%3Den%26user%3DjgWZfZ0AAAAJ%26citation_for_view%3DjgWZfZ0AAAAJ%3AY0pCki6q_DkC%26tzom%3D300.

271. *Social mating system and reproductive success in house wrens.* **Soukup, S. S. and Thompson, C. F.** 1, 1997, Behavioral Ecology, Vol. 9, pp. 43-48.

272. *Effects of extra-pair and within-pair reproductive success on the opportunity for selection in birds.* **Wittingham, L. A. and Dunn, P. O.** 1, 2005, Behavioral Ecology, Vol. 16, pp. 138–144.

273. **Willisch, C. S.** *The Ecology of Reproduction in Long-lived Male Alpine Ibex (Capra ibx): The Role of Age, Dominance, and Alternative Mating Tactics.* Neuchâtel : University of Neuchâtel, 2009. p. 110.

274. **Roughgarden, J.** *Evolution's Rainbow: Diversity, Gender, and Sexuality in Nature and People.* 1st ed. Berkeley : University of California Press, 2004. p. 474. 978-0520240735.

275. *Local smoothness and the price of anarchy in splittable congestion games.* **Roughgarden, T. and Schoppman, F.** 2015, Journal of Economic Theory, Vol. 156, pp. 317-342.

276. *Evolution of cooperation and skew under imperfect information.* **Akçay, E., et al.** 37, 2012, Proceedings of the National Academy of Sciences of the United States of America, Vol. 109, pp. 14936-14941.

277. **Roughgarden, J.** Incentivizing Biological Cooperation: Direct Benefits and the Economic Theory of the Firm. pp 227-244. [book auth.] S. B. Gissis, E. Lamm and A. Shavit. *Landscapes of Collectivity in the Life Sciences.* Cambridge : MIT Press, 2018, p. 432.
https://books.google.com/books?hl=en&lr=&id=SPpFDwAAQBAJ&oi=fnd&pg=PA227&dq=roughgarden+theory&ots=7-Sh4-oLBq&sig=BEg9eP2-MgvQ2paEDB5eEMQMuNE#v=onepage&q=roughgarden%20theory&f=false.

278. **Davies, A. P. C. and Shackelford, T. K.** Two human natures: Men and women evolved different psychologies. [book auth.] C. Crawford and D. Krebs. *Foundations of Evolutionary Psychology.* s.l. : Lawrence Erlbaum Associates, 2008.
https://www.amazon.com/Foundations-Evolutionary-Psychology-Charles-Crawford/dp/080585956X/ref=dp_ob_image_bk.

279. **Miller, G. F.** How mate choice shaped human nature: A review of sexual selection and human evolution. pp 43-87. [book auth.] C. Crawford and D. L. Krebs.

Handbook of Evolutionary Psychology: Ideas, Issues, and Applications. Mahwah : Lawrence Erlbaum Associates Inc., 1997, p. 680. https://www.amazon.com/Handbook-Evolutionary-Psychology-Issues-Applications/dp/0805816666.

280. *Nuptial gifts, resource budgets, and reproductive output in a polyandrous butterfly.* **Karlsson, B.** 8, 1998, Ecology, Vol. 79, pp. 2931-2940.

281. **Snook, R. R.** The Evolution of Polyandry pp 159-180. [book auth.] David Shuker and Leigh Simmons. *The Evolution of Insect Mating Systems.* Oxford : Oxford University Press, 2014, p. 352.

282. **Thornhill, R. and Alcock, J.** *The Evolution of Insect Mating Systems.* Oxford : Oxford Univeristy Press, 2000. p. 564. 978-1583484203.

283. *Female mate choice and the emergence of male coercion.* **Glover, K. M. and Crowley, P. H.** 12, 2017, Behavioral Ecology and Sociobiology, Vol. 71, p. 181.

284. *Sexual behaviour of male northern elephant seals: II. Female responses to potentially injurious encounters. Behaviour.* **Mesnick, S. L. and Le Boeuf, B. J.** 3-4, 1991, Behaviour, Vol. 117, pp. 262-280.

285. **Reiss, M and Sants, H.** *Behaviour and Social Organisation.* Cambridge : Cambridge University Press, 1987. p. 108. 9780521313285.

286. *Forced copulations and female contact guarding at a high male density in a calopterygidae damselfly.* **Cordero, A.** 1, 1999, Journal of Insect Behavior, Vol. 12, pp. 27-37.

287. *Male armaments and reproductive behavior in "nutcracker" camel crickets (Rhaphidophoridae, pristoceuthophilus).* **Conroy, L. P. and Gray, D. A.** 1, 2015, Insects, Vol. 6, pp. 85-99.

288. *Forced copulation as a conditional alternative strategy in camel crickets.* **Conroy, L. P. and Gray, D. A.** 9, 2014, Behavioral Ecology and Sociobiology, Vol. 68, pp. 1431–1439.

289. *American Naturalist.* **Allen, J. J., et al.** 1, 2017, Dramatic fighting by male cuttlefish for a female mate, Vol. 190, pp. 144-151.

290. *Male copulatory behavior interrupts Japanese flying squid Todarodes pacificus female spawning activity.* **Puneeta, P., et al.** 2016, Marine Ecology Progress Series, pp. 277-281.

291. *Beobachtungen bei einem Einbürgerungsversuch mit der Brautente (Lampronessa sponsa (L.)).* **Heinroth, O.** 1, 1910, Journal für Ornithologie, Vol. 58, pp. 101-156.

292. *Copulation behavior in Western Grebe (Aechmophorus occidentalis) and Clark's Grebe (Aechmophorus clarkii).* **Hayse, F. E. and Turner, D. G.** 2, 2017, Waterbirds, Vol. 40, pp. 168-172.

293. *Forced copulation in waterfowl.* **McKinney, F., Derrickson, S. R. and Mineau, P.** 3-4, 1983, Behaviour, Vol. 86, pp. 250-294.

294. *The limits of sexual conflict in the narrow sense: New insights from waterfowl biology.* **Brennan, P. L. and Prum, R. O.** 1600, 2012, Philosophical Transactions of the Royal Society of London B., Vol. 367, pp. 2324-2338.

295. *Forced copulation in captive mallards: II. Temporal factors.* **Cheng, K. M., Burnes, J. T. and McKinney, F.** 3, 1982, Animal Behaviour, Vol. 30, pp. 695- 699.

296. *Forced copulation in captive mallards III. Sperm competition.* **Cheng, K. M., Burnes, J. T. and McKinney, F.** 2, 1983, The Auk, Vol. 100, pp. 302-310.

297. *Forced copulation results in few extrapair fertilizations in Ross's and lesser snow geese.* **Dunn, P. O., et al.** 5, 1999, Animal Behaviour, Vol. 57, pp. 1071-1081.

298. **Bagemihl, Bruce.** *Biological Exuberance: Animal Homosexuality and Natural Diversity.* New York : Stonewall Inn Editions, 2000. p. 768. 978-0312253776.

299. *Coevolution of male and female genital morphology in waterfowl.* **Brennan, P. L R., et al.** 5, 2007, PLoS ONE, Vol. 2, p. e418.

300. *The evolution of genital shape variation in female cetaceans.* **Orbach, D. N., et al.** 2, 2017, Evolution, Vol. 72, pp. 261-273.

301. *Digest: Ontogenesis and evolutionary allometry shape divergent evolution of genitalia in female cetaceans.* **Wang, J. W. and Liao, W. B.** 2, 2018, Evolution, Vol. 72, pp. 404-405.

302. **Matsumura, Y., et al.** During mating in beetles: Stiffness gradient of the beetle penis facilitates propulsion in the spiraled female spermathecal duct. pp. 247-262. [ed.] Gorb E. Gorb S. *Functional Surfaces in Biology III: Diversity of the Physical Phenomena.* Cham : Springer, 2017, Vols. Series: Biologically-Inspired Systems (Book 10). Series: Biologically-Inspired Systems (Book 10).

303. *Risk factors for rape, physical assault, and posttraumatic stress disorder in women: Examination of differential multivariate relationships.* **Acierno, R., et al.** 6, 1999, Journal of Anxiety Disorders, Vol. 13, pp. 541-563.

304. *I said no: The impact of voicing non-consent on women's perceptions of and responses to rape.* **Cook, N. K. and Messman-Moore, T. L.** 5, 2017, Violence Against Women, Vol. 24, pp. 507-527.

305. **Edwards, K. M, Neal, A. M. and Rodenhizer-Stämpfli, K. A.** Domestic violence prevention. (pp. 215-227. [book auth.] B. Teasdale and M. S. Bradley. *Advances in prevention science. Preventing crime and violence.* 1st . s.l. : Springer, 2017, p. 388 .

306. **Edwards, K. M. and Sessarego, S. N.** Risk of rape across the ecosystem: Outlining a framework for sexual assault risk reduction and resistance education. pp 39-66. [book auth.] L. M. Orchowski and C. Gidycz. *Sexual Assault Risk Reduction and Resistance: Theory, Research, and Practice.* London : Academic Press, 2018, 3, p. 470.

307. *Psychoactive substances and violent offences: A retrospective analysis of presentations to an urban emergency department in Switzerland.* **Liakoni, E., et al.** 3, 2018, PLoS One, Vol. 13, p. e0195234.

308. *The epidemiology of physical attack and rape among crack-using women.* **Falck, R. S., et al.** 1, 2001, Violence and Victims, Vol. 16, pp. 79-89.

309. *Youth Risk Behavior Surveillance --- United States.* **Grunbaum, J. A., et al.** SS02, 2004, Morbidity and Mortality Weekly Report, Vol. 53, pp. 1-96. https://www.cdc.gov/mmwr/preview/mmwrhtml/mm5324a7.htm.

310. **Tjaden, P. and Thoennes, N.** *Full report of the prevalence, incidence, and consequences of violence against women: findings from the National Violence Against Women Survey.* Washington, D.C. : National Institute of Justice, 2000. p. 71. NCJ 183781.

311. *Microsatellite assessment of sneaked fertilizations and egg thievery in the fifteenspine stickleback.* **Jones, A. G., Östlund-Nilsson, S. and Avise, J. C.** 3, 2017, Evolution, Vol. 52, pp. 848-858.

312. *Male reproductive tactics in the threespine stickleback–an evaluation by DNA fingerprinting.* **Rico, C., Kuhnlein, U. and Fitzgerald, G. J.** 2, 1992, Molecular Ecology, Vol. 1, pp. 79-87.

313. *The evolution of bourgeois, parasitic, and cooperative reproductive behaviors in fishes.* **Taborsky, M.** 2, 2001, Journal of Heredity, Vol. 92, pp. 100–110.

314. *Post-mating clutch piracy in an amphibian.* **Vieites, D. R., et al.** 7006, 2004, Nature, Vol. 431, pp. 305-308.

315. **Larson, G.** *Unnatural Selections.* Kansas City : Andrews and McMeel, 1991. p. 107. 978-0836218817.

316. *Sneak male mating tactics between lampreys (Petromyzontiformes) exhibiting alternative life-history strategies.* **Hume, J. B., et al.** 3, 2013, FishBiology, Vol. 82, pp. 1093-1100.

317. *The rock–paper–scissors game and the evolution of alternative male strategies.* **Sinervo, B. and Lively, C. M.** 1996, Nature, Vol. 380, pp. 240-243.

318. *Relating endocrinology, physiology and behaviour using species with alternative mating strategies.* **Miles, D. B., et al.** 4, 2007, Functional Ecology, Vol. 21, pp. 653-665.

319. *Sexual selection and alternative mating behaviours generate demographic stochasticity in small populations.* **Calsbeek, R., et al.** 1487, 2001, Proceedings of the Royal Society B, Vol. 269.

320. *Mate choice games, context-dependent good genes, and genetic cycles in the side-blotched lizard, Uta stansburiana.* **Alonzo, S. H. and Sinervo, B.** 2-3, 2001, Behavioral Ecology and Sociobiology, Vol. 49, pp. 176–186.

321. *A dominant allele controls development into female mimic male and diminutive female ruffs.* **Lank, D. B., et al.** 6, 2013, Biology Letters, Vol. 9, p. 20130653. http://rsbl.royalsocietypublishing.org/content/9/6/20130653.

322. *Permanent female mimics in a lekking shorebird.* **Jukema, J. and Piersma, T.** 2, 2006, Biological Letters, Vol. 2, pp. 161-165.

323. *Assortative mating and divergent male courtship behaviours between two cryptic species of nine-spined sticklebacks (genus Pungitius).* **Meguro, Y., et al.** 15, 2016, Behaviour, Vol. 153, pp. 1879–1911.

324. *Homosexuality in the ten-spined stickleback (Pygosteus pungitius L.* **Morris, D.** 1, 1952, Behaviour, Vol. 4, pp. 233-261.

325. *Stealing fertilizations in the fourspine stickleback, Apeltes quadracus.* **Rowland, W. J.** 4, 1979, The American Naturalist, Vol. 114, pp. 602-604.

326. *Semen Displacement as a Sperm Competition Strategy in Humans.* **Gallup, G. G. Jr. and Burch, R. L.** 4, 2004, Evolution and Human Behavior, Vol. 24, pp. 277-289.

327. *Mate retention, semen displacement, and human sperm competition: a preliminary investigation of tactics to prevent and correct female infidelity.* **Goetz, A. T., et al.** 4, 2005, Personality and Individual Differences, Vol. 38, pp. 749-763.

328. *Psychological and physiological adaptations to sperm competition in humans.* **Shackelford, T. K., Nicholas, P. and Goetz, A. T.** 3, 2005, Review of General Psychology, Vol. 9, pp. 228-248.

329. *Mate retention, semen displacement, and human sperm competition: A preliminary investigation of tactics to prevent and correct female infidelity.* **Goetz, A. T., et al.** 4, 2005, Personality and Individual Differences, Vol. 38, pp. 749-763.

330. *Sperm competition in marriage: Semen displacement, male rivals, and spousal discrepancy in sexual interest.* **Pham, M. N., DeLecce, T. and Shackelford, T. K.** 15, 2017, Personality and Individual Differences, Vol. 105, pp. 229-232.

331. *Sperm removal during copulation confirmed in the oldest extant damselfly, Hemiphlebia mirabilis.* **Cordero-Rivera, A.** 2017, PeerJ, Vol. 4, p. e2077.

332. *Dual function of the damselfly penis: Sperm removal and transfer.* **Waage, J. K.** 4383, 1979, Science, Vol. 203, pp. 916–918.

333. *Sperm removal and ejaculate size correlate with chelae asymmetry in a freshwater crayfish species.* **Galeotti, P., et al.** 2008, Behavioral Ecology and Sociobiology, Vol. 62, pp. 1739–1745.

334. *Sperm removal, ejaculation and their behavioural interaction in male cuttlefish in response to female mating history.* **Wada, T., et al.** 3, 2010, Animal Behaviour, Vol. 79, pp. 613-619.

335. *Octapus senscence: The beginning of the end.* **Anderson, R. C., Wood, J. B. and Byrne, R. A.** 4, 2002, Journal of Applied Animal Welfare Science, Vol. 5, pp. 275-283.

336. *Resources, attractiveness, family commitment; Reproductive decisions in human mate choice.* **Bereczkei, T., et al.** 8, 1997, Ethology, Vol. 103, pp. 681-699.

337. *Old money, the nouveaux riches and Brunhilde's marriage strategy.* **Bronsert, A. K., Glazer, A. and Konrad, K. A.** 1, 2017, Journal of Population Economics, Vol. 30, pp. 163-186.

338. *Human mating strategies: from past causes to present consequences.* **Marzoli, D., Havlíček, J. and Roberts, S. C.** 2, 2018, Wiley Interdisciplinary Reviews: Cognitive Science, Vol. 9, p. e1456.

339. *Benefits of small-sized caches for scatter-hoarding rodents: Influence of cache size, depth, and soil moisture.* **Geluso, K.** 5, 2005, Journal of Mammology, Vol. 56, pp. 1186-1192.

340. *Counting on your friends: The role of social environment on quantity discrimination.* **McKeena-Kelly, E.** 2016, Behavioural Processes, Vol. 128, pp. 9-16.

341. *Can brown lemurs (Eulemur fulvus) learn to deceive a human competitor?* **Genty, E., Foltz, J. and Roder, J,-J.** 2, 2008, Animal Cognition, Vol. 11, pp. 255–266.

342. *Tactical deception to hide sexual behaviour: macaques use distance, not visibility.* **Overduin-deVries, A. M., et al.** 8, 2015, Behavioral Ecology and Sociobiology, Vol. 69, pp. 1333–1342.

343. *Factors influencing deceptive behaviours in Tonkean macaques (Macaca tonkeana).* **Canteloup, C., et al.** 7-8, 2017, Behaviour, Vol. 154, pp. 765–784.

344. **Trivers, R. L.** Parental investment and sexual selection. pp 136–179. [book auth.] B. Campbell. *Sexual Selection and the Descent of Man, 1871-1971.* Chicago : Aldine, 1972, p. 388.

345. **Vries, A. O.** Sneaky copulator. [book auth.] J. Vonk and T. Shackelford. *Encyclopedia of Animal Cognition and Behavior.* Cham : Springer, 2017. https://link.springer.com/referencework/10.1007%2F978-3-319-47829-6.

346. *Extra-pair paternity results from female preference for high-quality males in the blue tit.* **Kempenaers, B., et al.** 1992, Nature, Vol. 357, pp. 494–496. 1476-4687 (online).

347. *Requirements for a mixed reproductive strategy in avian species.* **Fitch, M. A. and Shugart, G. W.** 1, 1984, The American Naturalist, Vol. 124, pp. 116-126.

348. *The ecology and evolution of extra-pair copulation of birds.* **Westneat, D. F., Sherman, P. W. and Morton, M. L.** 1990, Current Ornithology, Vol. 7, pp. 331-369. http://web.as.uky.edu/biology/faculty/Westneat/Personal/Pubs%20Folder/EPCs1990.pdf.

349. *Extra-pair copulations in black-capped chickadees: The role of the female.* **Smith, S. M.** 1-2, 1988, Behaviour, Vol. 107, pp. 15-23.

350. *Extrapair fertilizations vary with female traits and pair composition, besides male attractiveness, in barn swallows.* **Costanzo, A., et al.** 2017, Animal Behaviour, Vol. 134, pp. 183-191.

351. *Female collared flycatchers choose neighbouring and older extra-pair partners from the pool of males around their nests.* **Edme, A., Munclinger, P. and Krist, M.** 4, 2016, Journal of Avian Biology, Vol. 47, pp. 552-562.

352. *ourtship calls in Blue Tits Cyanistes caeruleus: daily and seasonal occurrence and link to paternity.* **Schlicht, L. and Kempenaers, B.** 2, 2016, Ardea, Vol. 104, pp. 107-117.

353. *Gray fox home range, spatial overlap, mated pair interactions and extra-territorial forays in southwestern Georgia, USA.* **Deuel, N. R., et al.** 1, 2017, Wildlife Biology, Vol. 2018, pp. wlb-00326.

354. *Polygynandrous anoles and the myth of the passive female.* **Stamps, J.A.** 7, 2018, Behavioral Ecology and Sociobiology, Vol. 72, p. 107.

355. *Neighbor recognition in two "solitary" carnivores: The raccoon (Procyon lotor) and the red fox (Vulpes fulva).* **Barash, D. P.** 4153, 1974, Science, Vol. 185, pp. 794-796.

356. *Egg distribution, mate-guarding intensity and offspring characteristics in dragonflies (Odonata).* **Schenk, K., Suhling, F. and Martens, A.** 3, 2004, Animal Behaviour, Vol. 68, pp. 599-606.

357. *Evidence for widespread sperm displacement ability among Zygoptera (Odonata) and the means for predicting its presence.* **Waage, J. K.** 3, 1986, Biological Journal of the Linnean Society, Vol. 28, pp. 285-300.

358. *The adaptive significance of postcopulatory mate guarding in a dragonfly, Pachydiplax longipennis.* **Sherman, K.J.** 4, 2006, Animal Behaviour, Vol. 31, pp. 1107-1115.

359. *Kinship and familiarity mitigate costs of social conflict between Seychelles warbler neighbors.* **Bebbington, K., et al.** 2017, Proceedings of the National Academy of Sciences, p. 201704350.

360. *Familiarity with neighbours affects intrusion risk in territorial red squirrels.* **Siracusa, E., et al.** 2017, Animal Behaviour, Vol. 133, pp. 11-20.

361. *The effects of habitat deterioration and social status on patrolling behavior in the territorial damselfly Calopteryx splendens.* **Golab, M. J., et al.** 1, 2017, Polish Journal of Ecology, Vol. 65, pp. 122-131.

362. *Time budgets and territory size: Some simultaneous optimization models for energy maximizers.* **Schoener, T. W.** 2, 1987, American Zoologist, Vol. 27, pp. 259-291.

363. *Territorial defence and territory take-over by satellite males in the waterbuck Kobus ellipsiprymnus (Bovidae).* **Wirtz, P.** 2, 1981, Behavioral Ecology and Sociobiology, Vol. 8, pp. 161-162.

364. *Alternative male mating tactics in a cichlid, Pelvicachromis pulcher: A comparison of reproductive effort and success.* **Martin, E. and Taborsky, M.** 5, 1997, Behavioral Ecology and Sociobiology, Vol. 41, pp. 311-319.

365. *Male violence and sexual intimidation in a wild primate society.* **Baniel, A., Cowlishaw, G. and Huchard, E.** 14, 2017, Current Biology, Vol. 27, pp. 2163-2168.e3.

366. *Male aggression against women: An evolutionary perspective.* **Smuts, Barbara.** 1, 1992, Human Nature, Vol. 3, pp. 1–44.

367. *Baboons eavesdrop to deduce mating opportunities.* **Crockford, C., et al.** 5, 2007, Animal Behaviour, Vol. 73, pp. 885-890.

368. *Social eavesdropping and the evolution of conditional cooperation and cheating strategies.* **Earley, R. L.** 1553, 2010, Philosophical Transactions of the Royal Society B, Vol. 365, pp. 2675-2686.

369. **Swedell, L. and Schreier, A.** Male aggression towards females in hamadryas baboons: Conditioning, coercion, and control. pp 244–268. [book auth.] M. N. Muller and R. W. Wrangham. *Sexual Coercion in Primates and Humans: An Evolutionary Perspective on Male Aggression Against Females.* Cambridge : Harvard University Press, 2009, 18, p. 504.

370. *Dominance and reproduction in Baboons (Papio cynocephalus).* **Hausfater, G.** 1975, Contributions to Primatology, Vol. 7, pp. 1-150.

371. *Sexually coercive male chimpanzees sire more offspring.* **Feldblum, J. T., et al.** 23, 2014, Current Biology, Vol. 24, pp. P2855-2860.

372. *Information content of female copulation calls in yellow baboons.* **Semple, S., et al.** 1, 2002, American Journal of Primatology, Vol. 56, pp. 43-56.

373. *Owners and Satellites: The economics of territorial defence in the pied wagtail, Motacilla alba.* **Davies, N. B. and Houston, A. I.** 1, 1981, Journal of Animal Ecology, Vol. 50, pp. 157-180.

374. *Calling as an ownership convention on pied wagtail territories.* **Davies, N.B.** 2, 1981, Animal Behaviour, Vol. 29, pp. 529-534.

375. **Close, S. K. and Close-Stacey, K.** *Elderly Slaves of the Plantation South.* New York : Taylor & Frances Inc., 1996. p. 138. 978-0815327561.

376. **Morgan, P. D.** *Slave Counterpoint: Black culture in the Eighteenth-Century Chesapeake and Lowcountry.* 9th. Chapel Hill : University of North Carolina Press Books, 1998. p. 736. 978-0807824092.

377. **Conrad, R. E.** *Destruction of Brazilian Slavery, 1850-88.* Berkeley : University of California Press, 1972. p. 362. 978-0520021396.

378. **Barrow, R.H.** *Slavery in the Roman Empire.* facsimile reprin. s.l. : Kessinger Publihing Co., 2010/1968. p. 393. 978-1164498148.

379. *The reproductive behaviour of α-, β-, and γ-male morphs in Paracerceis sculpta, a marine isopod crustacean.* **Shuster, S. M.** 3-4, 1992, Behaviour, Vol. 121, pp. 231-258.

380. *Alternative reproductive behaviors: Three discrete male morphs in Paracerceis sculpta, an intertidal isopod from the northern Gulf of California.* **Shuster, S. M.** 2, 1987, Journal of Crustacean Biology, Vol. 7, pp. 318-327.

381. *Female mimicry and resource defense polygyny by males of a tropical rove beetle, Leistotrophus versicolor (Coleoptera : Staphylinidae).* **Forsyth, A. and Alcock, J.** 5, 1990, Behavioral Ecology and Sociobiology, Vol. 26, pp. 325-330.

382. *It pays to cheat: Tactical deception in a cephalopod social signalling system.* **Brown, C., Garwood, M. P. and Williamson, J. E.** 5, 2012, Biology Letters, Vol. 8, p. rsbl20120435.

383. *Behavioural ecology: transient sexual mimicry leads to fertilization.* **Hanlon, R. T., et al.** 7023, 2005, Nature, Vol. 433, p. 212.

384. *Tactic-specific success in sperm competition.* **Fu, W. N., Neff, B. D. and Gross, M. R.** 1472, 2001, Proceedings of the Royal Society of London, B, Vol. 268, pp. 1105-1112.

385. *Courtship and mating behaviour of the Chinese praying mantis, Tenodera aridifolia sinensis.* **Liske, E. and Davis, W. J.** 5, 1987, Animal Behaviour, Vol. 35, pp. 1524-1537.

386. *Cannibalism in the praying mantis Hierodula membranacea.* **Birkhead, T. R., Lee, K. E. and Young, P.** 1, 1988, Behaviour, Vol. 106, pp. 112-118.

387. *Fitness benefits of polyandry for experienced females.* **Whittingham, L. A. and Dunn, P. O.** 11, 2010, Molecular Ecology, Vol. 19, pp. 2328–2335.

388. **Rossner, J.** *Looking for Mr. Goodbar.* New York : Simon & Schuster, 2014/1975. p. 288. 978-1476774725.

389. *Plant defence signals and Batesian mimicry.* **Augner, M. and Bernays, E. A.** 6, 1998, Evolutionary Ecology, Vol. 12, pp. 667-679.

390. **Brown, H.G.** *Sex and the Single Girl.* New York : Barricade Books, 1962/2003. p. 267. 978-1569802526.

391. **Ginsberg, A.** Demonstration or Spectacle as Example, As Communication, or How to Make a March/Spectacle pp 208-212. [book auth.] A. Charles. *The Portable Sixties Reader.* Berkeley : Penguin Classic, 1965/2002, p. 672.

392. *Mechanisms and evolution of deceptive pollination in orchids.* **Jersáková, J., Johnson, S. D. and Kindlmann, P.** 2, 2006, Biological Reviews, Vol. 81, pp. 219-235.

393. *Evidence for Bastesian mimicry in a butterfly-pollinated orchid.* **Johnson, S. D.** 1, 1994, Biological Journal of the Linnean Society, Vol. 53, pp. 91–104.

394. *Convergent evolution of sexual deception via chromatic and achromatic contrast rather than colour mimicry.* **Gaskett, A. C., Endler, J. A. and Phillips, R. D.** 2, 2016, Evolutionary Ecology, Vol. 31, pp. 205–227.

395. *Canlılarda Mimikri.* **Kanisanli, M. and Durmuş, S. H.** 1995, Ekoloji, Vol. 14, pp. 45-46.

396. *Deception in plants: mimicry or perceptual exploitation?* **Schaefer, H. M. and Ruxton, G. D.** 12, 2009, Trends in Ecology and Evolution, Vol. 24, pp. 676-685.

397. *Disorder in convergent floral nanostructures enhances signalling to bees.* **Moyroud, E., et al.** 2017, Nature, Vol. 550, pp. 469–474.

398. *Orchid pollination by sexual deception: Pollinator perspectives.* **Gaskett, A. C.** 1, 2011, Biological Reviews, Vol. 86, pp. 33-75.

399. *Hybrid floral scent novelty drives pollinator shift in sexually deceptive orchids.* **Vereecken, N. J., Cozzolino, S. and Schiestl, F. P.** 2010, BMC Evolutionary Biology, Vol. 10, p. 103.

400. *The flower of seduction.* **Ledford, H.** 2007, Nature, Vol. 445, pp. 816-817.

401. *The chemistry of sexual deception in an orchid-wasp pollination system.* **Schiestl, F. P., et al.** 5644, 2003, Science, Vol. 302, pp. 437-438.

402. *Species-specific male pollinators found for three native New Zealand greenhood orchids (Pterostylis spp.) suggest pollination by sexual deception.* **Thalwitzer, L., et al.** 3, 2018, Australian Journal of Botany, Vol. 66, pp. 243-254.

403. *Orchid sexual deceit provokes ejaculation.* **Gaskett, A. C., Winnick, C. G. and Herberstein, M. E.** 2008, The American Naturalist, pp. E206-212.

404. *The evolution of imperfect floral mimicry.* **Vereecken, N. J. and Schiestl, F. P.** 21, 2008, Proceedings of the National Academy of Science, Vol. 105, pp. 7484-7488.

405. *How an orchid harms its pollinatozzzzz.* **Wong, B. B. M. and Schiestl, F. P.** 1500, 2002, Proceedings of the Royal Society of London B, Vol. 269, pp. 1592-1532.

406. *Responses of male Zaspilothynnus trilobatus Turner wasps to females and the sexually deceptive orchid it pollinates.* **Peakall, R.** 2, 1990, Functional Ecology, Vol. 4, pp. 159-167.

407. *Crossing-distance effects in Delphinium nelsonii: Outbreeding and inbreeding depression in progeny fitness.* **Waser, N. M. and Price, M. V.** 3, 1994, Evolution, Vol. 48, pp. 842-852.

408. *Fly pollination of Gorteria diffusa (Asteraceae), and possible mimetic function for dark spots on the capitulum.* **Johnson, S. D. and Midgley, J. J.** 4, 1997, American Journal of Botany, Vol. 84, pp. 429-436.

409. *Atypical pseudopollen-forming hairs in Maxillaria (Orchidaceae).* **Davies, K. L., Turner, M. P. and Gregg, A.** 2, 2003, Botanical Journal of the Linnean Society, Vol. 143, pp. 151–158.

410. *Deceptive orchids with Meliponini as pollinators.* **Roubik, D. W.** 1-4, 2000, Plant Systematics and Evolution, Vol. 222, pp. 271-279.

411. *On the success of a swindle: pollination by deception in orchids.* **Schiest, F. P.** 6, 2005, Naturwissenschaften, Vol. 92, pp. 255–264.

412. *Pollination: Rotting smell of dead-horse arum florets.* **Stensmyr, M. C., et al.** 2002, Nature, Vol. 420, pp. 625-626.

413. *Function of the heater: The dead horse arum revisited.* **Angioy, A. -M., et al.** 3, 2004, Proceedings of the Royal Society of London B, Vol. 271, pp. S13-S15.

414. *Why are there so many orchid species?* **Dodson, C. H.** 2, 2003, Lankesteriana, Vol. 3, pp. 99-103.

415. *Ants on swollen-thorn acacias: species coexist in a simple system.* **Young, T. P., Stubblefield, C. H. and Isabell, L. A.** 1, 1997, Oecologia, Vol. 109, pp. 98-107.

416. *Economy of scale: third partner strengthens a keystone ant-plant mutualism.* **Prior, K. M. and Palmer, T. M.** 2, 2018, Ecology, Vol. 99, pp. 335-346.

417. *Breakdown of an ant-plant mutualism follows the loss of large herbivores from an African savanna.* **Palmer, T. M., et al.** 5860, 2008, Science, Vol. 319, pp. 192- 195. http://www.sciencemag.org/cgi/pmidlookup?view=long&pmid=18187652.

418. *Symbiotic ants as an alternative defense against giraffe herbivory in spinescent Acacia drepanolobium.* **Madden, D. and Young, T. P.** 2, 1992, Oecologia, Vol. 91, pp. 235-238.

419. *Defensive plant-ants stabilize megaherbivore-driven landscape change in an African savanna.* **Goheen, J. R. and Palmer, T. M.** 19, 2010, Current Biology, Vol. 20, pp. 1768-1772.

420. *Defensive ant, aphid and caterpillar mimicry in plants?* **Lev-Yadun, S. and Inbar, M.** 3, 2002, Biological Journal of the Linnean Society, Vol. 77, pp. 393-398.

421. *Weapon (thorn) automimicry and mimicry of aposematic colorful thorns in plants.* **Lev-Yadun, S.** 2, 2003, Journal of Theoretical Biology, Vol. 224, pp. 183-188.

http://www.esalq.usp.br/lepse/imgs/conteudo_thumb/Weapon--thorn--automimicry-and-mimicry-of-aposematic-colorful-thorns-in-plants-1.pdf.

422. *The role of dark flowers in the pollination of certain Umbelliferae.* **Eisikowitch, D.** 5, 1980, Journal of Natural History, Vol. 14, pp. 737–742.

423. **Lev-Yadun, S.** Gall Midge Mimicry 323-325. [book auth.] Simcha Lev-Yadun. *Defensive (anti-herbivory) Coloration in Land Plants.* Cham : Springer, 2016, 62, p. 385. https://www.amazon.com/Defensive-anti-herbivory-Coloration-Land-Plants/dp/3319420941/ref=sr_1_1?s=books&ie=UTF8&qid=1533337918&sr=1-1&keywords=Defensive+%28anti-herbivory%29+Coloration+in+Land+Plants.

424. *Crop mimicry in weeds.* **Barrett, S. H.** 3, 1983, Economic Botany, Vol. 37, pp. 255–282.

425. *Mimicry in plants.* **Barrett, S. H.** 9, 1987, Scientific America, Vol. 255, pp. 76-83. http://labs.eeb.utoronto.ca/barrett/pdf/schb_54.pdf.

426. *The evolution of floral gigantism.* **Davis, C. C., Endress, P. K. and Baum, D. A.** 1, 2008, Current Opinion in Plant Biology, Vol. 11, pp. 49-57.

427. *Feeding enhances photosynthetic efficiency in the carnivorous pitcher plant Nepenthes talangensis.* **Pavlovic, A., et al.** 2, 2009, Annals of Botany, Vol. 104, pp. 307-314.

428. *New evidence on the origin of carnivorous plants.* **Givnish, T. J.** 1, 2015, Proceedings of the National Acadamey of Sciences of the United States of America, Vol. 112, pp. 10-11.

429. *Carnivorous plants: Phylogeny and structural evolution.* **Albert, V. A., Williams, S. E. and Chase, M. W.** 5076, 1992, Science, Vol. 257, pp. 1491-1495. https://www.ncbi.nlm.nih.gov/pubmed/1523408.

430. *Energetics and the evolution of carnivorous plants—Darwin's 'most wonderful plants in the world'.* **Ellison, A. M. and Gotelli, N. J.** 1, 2009, Journal of Experimental Botany, Vol. 60, pp. 19-42.

431. *Ultraviolet patterns in the traps of carnivorous plants.* **Joel, D. M., Juniper, B. E. and Dafni, A.** 1985, New Phytology, Vol. 101, pp. 585-593.

432. *Genome of the pitcher plant Cephalotus reveals genetic changes associated with carnivory.* **Fukushima, K., et al.** 0059, 2017, Nature Ecology and Evolution, Vol. 1, p. 0059. Sankoff, D.; Shibata, T. F.; Shigenobu, S.; Sumikawa, N.; Uzawa, T.; Xie, M.; Zheng, C.; Pollock, D. D.; Albert, V. A.; Li, S.; Hasebe, M..

433. *Carnivory in Byblis revisited II: the phenomenon of symbiosis on insect trapping plants.* **Hartmeyer, S.** 4, 1998, Carnivorous Plant Newsletter, Vol. 27, pp. 110–113.

434. *Review of the Australian stilt flies (Diptera : Micropezidae) with a phylogenetic analysis of the family.* **McAlpine, D. K.** 1, 1998, Invertebrate Taxonomy, Vol. 12, pp. 55-134.

435. *The coevolution of overconfidence and bluffing in the resource competition game.* **Li, K., et al.** 2016, Scientific Reports, Vol. 6, p. 21104.

436. *The trouble with overconfidence.* **Moore, D. A. and Healy, P. J.** 2, 2008, Psychology Review, Vol. 115, pp. 502-517.

437. *The causes of impasses in labor disputes.* **Babcock, L. C. and Olson, C. A.** 2, 1992, Industrial Relations: A Journal of Economy and Society, Vol. 31, pp. 348-360.

438. *Self-deception in an evolutionary game.* **Byrne, C. C. and Kurland, J. A.** 4, 2001, Journal of Theoretical Biology, Vol. 212, pp. 457-480.

439. *The evolution of misbelief.* **McKay, R. T. and Dennett, D. C.** 6, 2009, Behavioral and Brain Science, Vol. 32, pp. 493-510.

440. *The elements of a scientific theory of self-deceoption.* **Trivers, R.** 1, 2006, Annals of the New York Academy of Sciences, Vol. 907, pp. 114-131.

441. *The evolution and psychology of self-deception.* **von Hippel, W. and Trivers, R.** 1, 2011, The Behavioral and Brain Sciences, Vol. 34, pp. 1-16.

442. **Kusch, J. M.** *The Predictors and Consequences of Individual Variation in Sociality in Black-Tailed Prairie Dogs (Cynomys Ludovicianus).* Saskatoon : University ofSaskatchewan, 2018. p. 109.

443. *The evolution of alloparental care and adoption in mammals and birds.* **Riedman, M.L.** 4, 1982, Quarterly Review of Biology, Vol. 57, pp. 405-435.

444. *Alloparental feeding in the king penguin.* **Lecomte, N., et al.** 2, 2006, Animal Behaviour, Vol. 71, pp. 457-462.

445. *King penguin brooding and defending a subAntarctic skua chick.* **Oosthuizen, W. C. and de Bruyn, N.** 2, 2009, Polar Biology, Vol. 32, pp. 303-305.

446. *The process of domestication.* **Clutton-Brock, J.** 2, 1992, Mammal Review, Vol. 22, pp. 79–85.

447. *Naturalizing Anthropomorphism: Behavioral Prompts to Our Humanizing of Animals.* **Horowitz, A. C. and Bekoff, M.** 1, 2007, Anthrozoos: A Multidisciplinary Journal of The Interactions of People & Animals, Vol. 20, pp. 23-35.

448. *On cuteness: Unlocking the parental brain and beyond.* **Kringelbach, M. L., et al.** 7, 2016, Trends in Cognitive Sciences, Vol. 20, pp. P545-55.

449. *Viewing cute images increases behavioral carefulness.* **Sherman, G. D., Haidt, J. and Coan, J. A.** 2, 2009, Emotion, Vol. 9, pp. 282-286.

450. *Studien zur Vererbung von Verhaltensweisen: 3. Verhalten gegenüber jungen Katzen.* **Spindler, P.** 1, 1961, Anthropologischer Anzeiger, Vol. 25, pp. 60-80.

451. **McCabe, V.** Facial proportions, perceived age, and caregiving. pp. 89–95. [book auth.] T. R. Alley. *Social and Applied Aspects of Perceiving Faces.* Hillsdale : Erlbaum, 1988, p. 304. https://www.amazon.com/Applied-Perceiving-Resources-Ecological-Psychology/dp/0805801634/ref=sr_1_1?s=books&ie=UTF8&qid=1541283861&sr=1-1&keywords=Social+and+Applied+Aspects+of+Perceiving+Faces.

452. *Baby talk to the babyfaced.* **Zebrowitz, L. A., Brownlow, S. and Olson, K.** 3, 1992, Journal of Nonverbal Behavior, Vol. 16, pp. 143–158.

453. *Female facial beauty: The fertility hypothesis.* **Johnston, V. S.** 1, 2000, Pragmatics & Cognition, Vol. 8, pp. 107-122.

454. *Is beauty in the eye of the beholder?* **Johnston, V. S. and Franklin, M.** 3, 1993, Ethology and Sociobiology, Vol. 14, pp. 183-199.

455. *The significance of beauty.* **Dunlop, K.** 3, 1918, Psychological Review, Vol. 25, pp. 171-255.

456. *The evolutionary psychology of facial beauty.* **Rhodes, G.** 2006, Annual Review of Psychology, Vol. 57, pp. 199-226.

457. *Sexual selection, physical attractiveness, and facial neoteny.* **Jones, D.** 5, 1995, Current Anthropology, Vol. 36, pp. 723–748.

458. *Criteria of facial attractiveness in five populations.* **Jones, D. and Hill, K.** 3, 1993, Human Nature, Vol. 4, pp. 271-296.

459. *Do women's mate preferences change across the ovulatory cycle? A meta-analytic review.* **Gildersleeve, K., Haselton, M. G. and Fales, M. R.** 5, 2014, Psychological Bulletin, Vol. 140, pp. 1205-1259.

460. *Menstrual cycle alters face preference.* **Penton-Voak, I. S., et al.** 6738, 1999, Nature, Vol. 399, pp. 741-742.

461. *Women's emotional and sexual attraction to men across the menstrual cycle.* **Shimoda, R., Campbell, A. and Barton, R. A.** 1, 2018, Behavioral Ecology, Vol. 29, pp. 51–59.

462. *Women's preference for dominant male odour: effects of menstrual cycle and relationship status.* **Havlicek, J., Roberts, S. C. and Flegr, J.** 3, 2005, Biological Letters, Vol. 1, pp. 256-259.

463. *Commitment to relationships and preferences for femininity and apparent health in faces are strongest on days of the menstrual cycle when progesterone level is high.* **Jones, B. C., et al.** 3, 2005, Hormones and Behavior, Vol. 48, pp. 283-290.

464. *Corrigendum to "Commitment to relationships and preferences for femininity and apparent health in faces are strongest on days of the menstrual cycle when progesterone level is high" [Horm. Behav. 48 (2005) 283–290].* **Jones, B. C., et al.** 3, 2014, Hormones and Behavior, Vol. 66, p. 567.

465. *Paternally inherited HLA alleles are associated with women's choice of male odor.* **Jacob, S., et al.** 2, 2002, Nature Genetics, Vol. 30, pp. 175-179.

466. *Body odour preferences in men and women: do they aim for specific MHC combinations or simply heterozygosity?* **Wedekind, C. and Füri, S.** 1387, 1997, Proceedings of the Royal Society B, Vol. 264, pp. 1471–1479.

467. *MHC-dependent mate preference in humans.* **Wedekind, C., et al.** 1359, 1995, Proceedings of the Royal Society B, Vol. 260, p. Proceedings of the Royal Society B.

468. *Accumulating evidence suggests that men do not find body odours of human leucocyte antigen-dissimilar women more attractive.* **Lobmaier, J. S., et al.** 1878, 2018, Proceedings of the Royal Society B, Vol. 285, p. 20180566.

469. *Men's preferences for women's body odours are not associated with human leucocyte antigen.* **Probst, F., et al.** 1864, 2017, Proceedings of the Royal Society B, Vol. 284, p. 20171830.

470. **Gould, S. J.** *Ontogeny and Phylogeny.* Cambridge : Belknap Press, 1985. p. 520. 978-0674639416.

471. **Campbell, Bernard.** *Sexual Selection and the Descent of Man: The Darwinian Pivot.* s.l. : Transaction Publishers, 1972. p. 388. 978-0202020051.

472. *Punishing pregnant drug-using women: defying law, medicine and common sense.* **Flavin, J. and Paltrow, L. M.** 2, 2010, Journal of Addictive Diseases, Vol. 29, pp. 231-244.

473. *Initial symptoms of nicotine dependence in adolescents.* **DiFranza, J. R., et al.** 3, 2000, Tobacco Control, Vol. 9, pp. 313-319.

474. *Initial exposure to nicotine in college-age women smokers and never-smokers: a replication and extension.* **Pomerleau, C. S., et al.** 3, 1999, Journal of Addictive Diseases, Vol. 18, pp. 13-19.

475. *Prescription opioid analgesics rapidly change the human brain.* **Younger, J. W., et al.** 8, 2011, Pain, Vol. 152, pp. 1803-1810.

476. *Intrauterine exposure to tobacco and executive functioning in high school.* **Rose-Jacobs, R., et al.** 1, 2017, Drug and Alcohol Devpendency, Vol. 176, pp. 169-175.

477. *Alcohol and aldehyde dehydrogenase polymorphisms and the risk for alcoholism.* **Higuchi, S., et al.** 8, 1995, American Journal of Psychiatry, Vol. 152, pp. 1219-1221.

478. *Genetic contributors to variation in alcohol consumption vary by race/ethnicity in a large multi-ethnic genome-wide association study.* **Jorgenson, E., et al.** 2017, Molecular Psychiatry, Vol. 22, pp. 1359–1367.

479. *Associations of ADH and ALDH2 gene variation with self report alcohol reactions, consumption and dependence: an integrated analysis.* **Macgregor, S., et al.** 3, 2009, Human Molecular Genetics, Vol. 18, pp. 580-593.

480. *Genetic and environmental contributions to alcohol abuse and dependence in a population-based sample of male twins.* **Prescott, C. A. and Kendler, K. S.** 1, 1999, American Journal of Psychiatry, Vol. 156, pp. 34-40.

481. *Chapter 48 - Neuroepigenetics and addiction.* **Walker, D. M. and Nestler, E. J.** 2018, Handbook of Clinical Neurology, Vol. 148, pp. 747-765.

482. *Moderation of the effect of adolescent-onset cannabis use on adult psychosis by a functional polymorphism in the catechol-o-methyltransferase gene: Longitudinal evidence of a gene X environment interaction.* **Caspi, A., et al.** 10, 2005, Biological Psychiatry, Vol. 57, pp. 1117-1127.

483. *Polysubstance abuse vulnerability genes: genome scans for association, using 1,004 subjects and 1,494 single-nucleotide polymorphisms.* **Uhl, G. R., et al.** 6, 2001, The American Journal of Human Genetics, Vol. 69, pp. 1290-1300.

484. *Alternatives to the War on Drugs.* **Alexander, B. K.** 1, 1990, Journal of Drug Issues, Vol. 20, pp. 1-27.

485. *Bureaucracy and morality.* **Dickson, D. T.** 2, 1968, Social Problems, Vol. 16, pp. 143-156.

486. *Creating visual differences: Methamphetamine users perceptions of anti-meth campaigns.* **Marsh, W., Copes, H. and Linnemann, T.** 1, 2017, International Journal of Drug Policy, Vol. 39, pp. 52-61.

487. *Can scare tactics and fear-based messages help deter substance misuse: A systematic review of recent (2005–2017) research.* **Esrick, J., et al.** 2018, Drugs: Education, Prevention and Policy, pp. 1-10.

488. **Reinarman, C. and Levine, H. G.** *Crack in America: Demon Drugs and Social Justice.* Berkeley : University of California Press, 1997. p. 359. 9780520202429.

489. **Festinger, L.** *A Theory of Cognitive Dissonance.* Anniversary edition. Stanford : Stanford University Press, 1957. p. 239 . 978-0804709118.

490. *The Influence of ambivalence on adolescents' reactions to anti-drug messages.* **Zhao, X. and Cappella, J. N.** 2, 2008, Communication Quarterly, Vol. 56, pp. 131-148.

491. *Peers and adolescent smoking.* **Kobus, K.** 1, 2003, Addiction, Vol. 98, pp. 37–55.

492. *Age differences in resistance to peer influence.* **Steinberg, L. and Monahan, K.** 6, 2007, Developmental Psychology, Vol. 43, pp. 1531-1543.

493. *Pre-adolescent receptivity to tobacco marketing and its relationship to acquiring friends who smoke and cigarette smoking initiation.* **Strong, D. R., et al.** 5, 2017, Annals of Behavioral Medicine, Vol. 51, pp. 730–740.

494. *The negro-to-black conversion experience.* **Cross, W. E. Jr.** 9, 1971, Black World, Vol. 20, pp. 13-27.

495. *Measuring nigrescence attitudes in school-aged adolescents.* **Gardner-Kitt, D. L. and Worrell, F. C.** 2, 2006, Journal of Adolescence, Vol. 30, pp. 187-202.

496. *Construct validity study of the Racial Identity Attitude Scale.* **Ponterotto, J. G. and Wise, S. L.** 2, 1987, Journal of Counseling Psychology, Vol. 34, pp. 218-223.

497. **Stevens, J. W.** African American female adolescent identity development: A three-dimensional perspective. pp 141-168. [book auth.] S. Jackson and S. Brissett-Chapman. *Serving African American Children.* s.l. : Routledge, 2018, 8, p. 274.

498. **Renn, O. and Levine, D.** Credibility and trust in risk communication. pp. 175-219. [book auth.] Roger E. Kasperson and Pieter Jan M. Stallen. *Communicating Risks to the Public: International Perspectives.* s.l. : Springer,, 1991, p. 481.

499. *How effective is drug abuse resistance education? A meta-analysis of Project DARE outcome evaluations.* **Ennett, S. T., et al.** 9, 1994, American Journal of Public Health, Vol. 84, pp. 1394-1401.

500. *Effects of hot spices on energy intake, appetite and sensory specific desires in humans.* **Reinbach, H. C., Martinussen, T. and Møller, P.** 6, 2010, Food Quality and Preferences, Vol. 21, pp. 655-661.

501. *The neurobiology of love.* **Esch, T. and Stefano, G. B.** 3, 2005, Neuroendocrinology Letters, Vol. 26, pp. 175-192.

502. *Ethnopharmacology of love.* **Leonti, M. and Casu, L.** 2018, Frontiers in Pharmacology, Vol. 9, p. 567.

503. **Butcher, J.** The concept of giving in Mexico, pp. 105-121. *Mexican Solidarity: Citizen Participation and Volunteering.* s.l. : Springer, 2010, p. 236 .

504. *Close social associations in animals and humans: functions and mechanisms of friendship.* **Massen, J. J. M., Sterck, E. H. M. and de Vos, H.** 11, 2010, Behaviour, Vol. 147, pp. 1379-1412.

505. *Endogenous pain control systems: Brainstem spinal pathways and endorphin circuitry.* **Basbaum, A. I. and Fields, H. L.** 1984, Annual Review of Neuroscience, Vol. 7, pp. 309-338.

506. *Endogenous and exogenous opioids in pain.* **Corder, G., et al.** 2018, Annual Review of Neuroscience , Vol. 41, pp. 453-473.

507. *Chromatin remodeling Iis a key mechanism underlying cocaine-induced plasticity in striatum.* **Kumar, A., et al.** 2, 2005, Neuron, Vol. 48, pp. 303-314.

508. *Transcription factor E2F3a in nucleus accumbens affects cocaine action via transcription and alternative splicing.* **Cates, Hannah M., et al.** 3, 2018, Biological Psychiatry, Vol. 84, pp. 167-179.

509. *Cocaine Cues and Dopamine in Dorsal Striatum: Mechanism of Craving in Cocaine Addiction.* **Volkow, N. D., et al.** 24, 2006, The Journal of Neuroscience, Vol. 26, pp. 6583-6588.

510. *Epigenetic mechanisms in drug addiction.* **Renthal, W. and Nestler, E. J.** 8, 2008, Trends in Molecular Medicine, Vol. 13, pp. 341-350.

511. *Triple reuptake inhibitors: The next generation of antidepressants.* **Marks, D., Pae, C. and Patkar, A.** 4, 2008, Current Neuropharmacology, Vol. 6, pp. 338–343.

512. *ncovering multiple molecular targets for caffeine using a drug target validation strategy combining A2A receptor knockout mice with microarray profiling.* **Yu, L., et al.** 3, 2009, Physiological Genomics, Vol. 37, pp. 199–210.

513. *Caffeine and the central nervous system: mechanisms of action, biochemical, metabolic and psychostimulant effects.* **Nehlig, A., Daval, J. -L. and Debry, G.** 2, 1992, Brain Research Reviews, Vol. 17, pp. 139-170.

514. **Davies, N.B.** *Dunnock Behaviour and Social Evolution.* New York : Oxford University Press, 1992. p. 288. 978-0198546740.

515. *The prairie vole: an emerging model organism for understanding the social brain.* **McGraw, L. A. and Young, L. J.** 2, 2010, Trends in Neurosciences, Vol. 33, pp. 103-109.

516. *Girls, aggression, and emotional regulation.* **Conway, A. M.** 2, 2005, Journal of Orthopsychiatry, Vol. 75, pp. 334-339.

517. *Aggress to impress: Hostility as an evolved context-dependent strategy.* **Griskevicus, V., et al.** 5, 2009, Journal of Personality and Social Psychology, Vol. 96, pp. 980-994.

518. *"Guess what I just heard!" Indirect aggression among teenage girls in Australia.* **Owens, L., Shute, R. and Slee, P.** 1, 2000, Aggressive Behavior, Vol. 26, pp. 67-83.

519. **Stewart, E. C. and Bennett, M. J.** *American Cultural Patterns: a Cross-Cultural Perspective.* London : Intercultural Press, Nicholas Brealey Publishing, 1972/1991. p. 228. 978-1877864018.

520. *The relationship between an innovation orientation and competitive strategies.* **Dobni, C. B.** 2, 2010, International Journal of Innovation Management, Vol. 14, pp. 331-357.

521. *Strategy shift: Integrating strategy and the firm's capability to innovate.* **Dobni, C. B. and Sand, C.** 5, 2018, Business Horizons, Vol. 61, pp. 797-808.

522. **Cukor, George.** *Why Can't a Woman Be More Like a Man?* [writ.] Alan Jay Lerner. [perf.] Rex Harrison. Warner Bros. Pictures, Inc, 1956.

523. *Sex Differences in Response to Visual Sexual Stimuli: A Review.* **Rupp, H. A. and Wallen, K.** 2, 2008, Archives of Sexual Behavior, Vol. 37, pp. 206–218.

524. *Sex-Specific Content Preferences for Visual Sexual Stimuli.* **Rupp, H. A. and Wallen, K.** 3, 2009a, Archives of Sexual Behaviour, Vol. 38, pp. 417–426.

525. *Menstrual cycle alters face preference.* **Penton-Voak, I. S., et al.** 1999, Nature, Vol. 399, pp. 741-742.

526. *Women's interest in visual sexual stimuli varies with menstrual cycle phase at first exposure and predicts later interest.* **Rupp, H. A. and Wallen, K.** 2, 2009b, Hormones and Behavior, Vol. 57, pp. 263-268.

527. *Men report stronger attraction to femininity in women's faces when their testosterone levels are high.* **Welling, L. L. M., et al.** 5, 2008, Hormones and Behavior, Vol. 54, pp. 703-708.

528. *The biology of human sexuality: Evolution, ecology and physiology.* **Bateman, P. W. and Bennett, N. C.** 1, 2006, Verbum Et Ecclesia, Vol. 27, pp. 245-264.

529. *Serial monogamy increases reproductive success in men but not in women.* **Jokela, M., et al.** 5, 2010, Behavioural Ecology, Vol. 21, pp. 906-912.

530. *HIV/AIDS, sexual and reproductive health: Intersections and implications for national programmes.* **Berer, M.** suppl 1, 2004, Health Policy and Planning, Vol. 19, pp. i62-i70.

531. *Sexuality — The Missing Link in Women' Health.* **Gordon, G. and Kanstrup, C.** 1, 1992, Institute of Developmental Studies Bulletin, Vol. 23, pp. 29–37.

532. *The evolution of infidelity in socially monogamous passerines: neglected components of direct and indirect selection.* **Griffith, S. C.** 2, 2007, American Naturalist, Vol. 169, pp. 274-281.

533. *Extra pair paternity in birds: a review of interspecific variation and adaptive function.* **Griffith, S. C., Owens, I. P. F. and Thuman, K. A.** 11, 2002, Molecular Ecology, Vol. 11, pp. 2195-2212.

534. *Monogamy in mammals.* **Kleiman, D.** 1, 1977, Quarterly Review of Biology, Vol. 52, pp. 39-69.

535. *Prairie-vole partnerships.* **Getz, L. L. and Carter, C. S.** 1, 1996, American Scientists, Vol. 84, pp. 56-62.

536. *Social behavior, reproduction and demography of the prairie vole, Microtus ochrogaster.* **Getz, L. L., McGuire, B. and Carter, C. S.** 2, 2003, Ethology, Ecology and Evolution, Vol. 15, pp. 105-118.

537. *Nucleus accumbens core medium spiny neuron electrophysiological properties and partner preference behavior in the adult male prairie vole, Microtus ochrogaster.* **Willett, J. A., et al.** 4, 2018, Journal of Neurophysiology, Vol. 119, pp. 1576-1588.

538. *Oxytocin Increases Generosity in Humans.* **Zak, P. J., Stanton, A. A. and Ahmadi, S.** 11, 2007, PLoS ONE, Vol. 2, p. e1128.

539. *Ventral striatopallidal oxytocin and vasopressin V1a receptors in the monogamous prairie vole (Microtus ochrogaster).* **Lim, M. M., Murphy, A. Z. and Young, I. J.** 4, 2004, Journal of Comparative Neurology, Vol. 468, pp. 555-580.

540. *Male-related pheromones and the activation of female reproduction in the prairie vole (Microtus ochrogaster).* **Carter, C. S., et al.** 5, 1980, Biology of Reproduction, Vol. 23, pp. 1038-1045.

541. *The effect of weanling and adult males on sexual maturation in female voles (Microtus ochrogaster).* **Hasler, M. J. and Nalbandov, A. V.** 2, 1974, General and Comparative Endocrinology, Vol. 23, pp. 237-238.

542. *Oxytocin, vasopressin, and the neurogenetics of sociality. (Review).* **Donaldson, Z. R. and Young, L. J.** 5903, 2008, Science, Vol. 322, pp. 900–904.

543. *The neural mechanisms and circuitry of the pair bond.* **Walum, H. and Young, L. J.** 2018, Nature Reviews: Neuroscience, Vol. 19, pp. 643–65.

544. *Cellular mechanisms of social attachment.* **Young, L. J., et al.** 2, 2001, Hormones and Behavior, Vol. 40, pp. 133-138.

545. *Oxytocin increases generosity in humans.* **Zak, P. J., Stanton, A. A. and Ahmadi, S.** 11, 2007, PLoS One, Vol. 2, p. e1128.

546. *Why do females mate multiply? A review of the genetic benefits.* **Jennions, M. D. and Petrie, M.** 1, 2000, Biological Reviews of the Cambridge Philosophical Society, Vol. 75, pp. 21-64.

547. *Male age, germline mutations and the benefits of polyandry.* **Radwan, J.** 7, 2003, Ecology Letters, Vol. 6, pp. 581-586.

548. *Women and aids in developing countries: The XIIth international conference on the social sciences and medicine.* **de Bruyn, M.** 3, 1992, Social Science and Medicine, Vol. 34, pp. 249-262.

549. *Exploring Sexual Terminology in a Vernacular in Rural Uganda.* **Ssali, A., et al.** Amsterdam : NL, 1997. Rural Uganda - Lessons for Health Education, paper presented at VIII International Conference on AIDS; Session 00255/D63.

550. *Is there a gender difference in strength of sex drive? Theoretical views, conceptual distinctions, and a review of relevant evidence.* **Baumeister, R. F., Catanese, K. R. and Vohs, K. D.** 3, 2001, Personality and Social Psychology Review, Vol. 5, pp. 242-273.

551. **Laumann, E. O., et al.** The Number of Partners, pp 172-222. *The Social Organization of Sexuality: Sexual Practices in the United States.* Chicago : University of Chicago Press, 2000, 5, p. 750.

552. **Standing, H. and Kisekka, M. N.** *Sexual behaviour in sub-Saharan Africa: a review and annotated bibliography.* Glasgow : Overseas Development Administration, 1989. p. 250. 0902500368.

553. *Adult sexual behavior in 1989: number of partners, frequency of intercourse and risk of AIDS.* **Smith, T. W.** 3, 1991, Family Planning Perspectives, Vol. 23, pp. 102-107.

554. *Sexual behaviours and contraception among university students in Turkey.* **Aras, S., et al.** 1, 2007, Journal of Biosocial Science,, Vol. 39, pp. 121-135.

555. *Gender differences in sexual behavior and condom-related behaviours and attitudes among Korean youths.* **Sohn, A. and Chun, S. S.** 2, 2007, Asia-Pacific Journal of Public Health, Vol. 19, pp. 45-52.

556. *Levels of Sexual Activity Among Adolescent Males in the United States.* **Sonenstein, F. L., Pleck, J. H. and Ku, L. C.** 4, 1991, Family Planning Perspectives, Vol. 23, pp. 162-167.

557. *Sexual behaviour of lesbians and bisexual women.* **Bailey, J. V., et al.** 2003, Sexually Transmitted Infections, Vol. 79, pp. 147-150.

558. *The web of human sexual contacts.* **Liljeros, F., et al.** 6840, 2001, Nature, Vol. 411, pp. 907-908.

559. *Prostitution and the sex discrepancy in reported number of sexual partners.* **Brewer, D. D., et al.** 22, 2000, Proceedings of the National Academy of Science USA, Vol. 97, pp. 12385-12388.

560. *Sexual behaviour changes among male sexual contacts of men with HIV disease: A 3 year overview.* **Calzavara, L., et al.** 1989. International Conference on AIDS. p. 733.

561. **Demissie, Z., et al.** *Characteristics of Health Programs.* U.S. Department of Health and Human Services Centers for Disease Control and Prevention . Atlanta : Center for Disease Control, 2014. p. 197.

562. *HIV education and sexual risk behaviors among young men who have sex with men.* **Raifman, J., Beyer, C. and Arrington-Sanders, R.** 2, 2018, LGBT Health, Vol. 5, pp. 131-138.

563. **Wright, J.** *Get Well Soon: History's Worst Plagues and the Heroes Who Fought Them.* New York : Henry Holt and Co, 2017. p. 336. 978-1627797467.

564. **Freud, A.** *The Ego and The Mechanisms of Defense.* Milton Park : Routledge, 1936/1992. p. 204 . 978-1855750388.

565. *Psychological stress and neuroendocrine function in humans: The last two decades of research.* **Biondi, M. and Picardi, A.** 3, 1999, Psychotherapy and Psychosomatics, Vol. 68, pp. 114–150.

566. *Endocrine, cardiovascular, and psychological correlated of olfactory sensitivity changes during the human menstrual cycle.* **Doty, R. L., et al.** 1, 1981, Journal of Comparative and Physiological Psychology, Vol. 95, pp. 45-60.

567. *Stress and hormones.* **Ranabir, S. and Reetu, K.** 1, 2011, Indian Journal of Endocrinology and Metabolism, Vol. 15, pp. 18–22.

568. *Endocrine response patterns and psychological correlates.* **Vaernes, R., et al.** 2, 1982, Journal of Psychosomatic Research, Vol. 26, pp. 123-131.

569. *Suppressing unwanted memories by executive control.* **Anderson, M. C. and Green, C.** 6826, 2001, Nature, Vol. 410, pp. 366-369.

570. *Region-based approach versus mechanism-based approach to the brain.* **Northoff, G.** 2, 2010, Neuropsychoanalysis: An Interdisciplinary Journal for Psychoanalysis and the Neurosciences, Vol. 12, pp. 167-170.

571. *In self-defense: Disruptions in the sense of self, lateralization and primitive defenses.* **Salas, C. and Turnbull, O. H.** 2, 2010, Neuropsychoanalysis: An Interdisciplinary Journal for Psychoanalysis and the Neurosciences, Vol. 12, pp. 172-182.

572. *Overcoming fixation with repeated memory suppression.* **Angello, G., Storm, B. C. and Smith, S. M.** 3, 2015, Memory, Vol. 23, pp. 381-389.

573. *Suppressing unwanted memories reduces their unintended influences.* **Hu, X., et al.** 2, 2017, Directions in Psychological Science, Vol. 26, pp. 197-206.

574. *Optimal foraging, the marginal value theorem.* **Charnov, E. L.** 2, 1976, Theoretical Population Biology, Vol. 9, pp. 129-136.

575. *Resource scarcity, effort, and performance in physically demanding jobs: An evolutionary explanation.* **Marko, P. and Stefan, T.** 3, 2018, Journal of Applied Psychology, Vol. 103, pp. 237-248.

576. **Kruuk, H.** *Predators and Anti-predator Behaviour of the Black-headed Gull:(Laurus Ridibundus L.).* s.l. : Brill Archive, 1964. pp. 1-129. Vol. 11, Behaviour Supplements 11; ASIN: B0006CL240; https://www.amazon.com/Predators-anti-predator-behaviour-black-headed-ridibundus/dp/B0006CL240/ref=sr_1_fkmr0_1?ie=UTF8&qid=1544302647&sr=8-1-fkmr0&keywords=Predators+and+Anti-predator+Behaviour+of+the+Black-h.

577. *Predation risk drives the expression of mobbing across bird species.* **da Cunha, F. C. R., Fontenelle, J. C. R. and Griesser, M.** 6, 2017, Behavioral Ecology, Vol. 28, pp. 1517-1523.

578. *The evolution of maladaptation.* **Crespi, B. J.** 6, 2000, Heredity, Vol. 84, pp. 623-629.

579. **Nash, J. F.** *Non-cooperative Games (Doctoral Dissertation).* Prinston : Princeton University, 1951. p. 27.

580. **National Retail Federation.** National Retail Security Survey 2018. *National Retail Security Survey 2018.* [Online] June 6, 2018. [Cited: 12 8, 2018.] https://nrf.com/research/national-retail-security-survey-2018.

581. **Hollinger, R. C. and Adams, A.** *2010 National Retail Security Survey. University of Florida, Gaineville.* Gainesville : Department of Sociology and Center for Studies in Criminology and Law, 2011. p. 28.

582. **Finklea, Kristin M.** *Organized Retail Crime.* Washington D.C. : Congressional Research Service Report for Congress, 2012. p. 26. R41118 .

583. **Simon, J.** *Governing Through Crime: How the War on Crime Transformed American Democracy and Created a Culture of Fear.* New York : Oxford University Press, 2007. p. 344. 978-0195386011.

584. *Social hierarchy: the self-reinforcing nature of power and status.* **Maggie, J. C. and Galinsky, A. D.** 1, 2008, The Academy of Management Annals, Vol. 2, pp. 351-398.

585. **Volpato, C.** *Deumanizzazione: Come si legittima la violenza (Universale Laterza Vol. 919).* Bari : Gius, Laterza & Figli, 2011. 978-8842096047.

586. *Formally shaming white-collar criminals.* **Ivancevich, J. M., Konopaske, R. and Gilbert, J. A.** 5, 2008, Business Horizons, Vol. 51, pp. 401-310.

587. *Do perceptions of punishment vary between white-collar and street crimes?* **Schoepfer, A., Carmichael, S. and Piquero, N. L.** 2, 2007, Journal of Criminal Justice, Vol. 35, pp. 151-163.

588. **Lewis, M.** *The Big Short: Inside the Doomsday Machine.* New York : W. W. Norton and Company, 2011. p. 291. 978-0393338829.

589. *Discourses of dehumanization: Enemy construction and canadian media complicity in the framing of the war on terror.* **Steuter, E. and Wills, D.** 2, 2009, Global Media Journal -- Canadian Edition, Vol. 2, pp. 7-24.

590. *The role of labeling and bias in the portrayals of acts of "terrorism": Media representations of muslims vs. non-Muslims.* **West, K. and Lloyd, J.** 2, 2017, Journal of Muslim Minority Affairs, Vol. 37, pp. 211-222.

591. *Framing analysis, dramatism and terrorism coverage: Politician and press responses to the Madrid airport bombing.* **Canel, M. J. and Garcia-Gurrionero, M.** 4, 2016, Communication and Society, Vol. 29, pp. 133-149.

592. *Meanings of Integration in the Australian press coverage of Muslims: Implications for social inclusion and exclusion.* **Rane, H. and Hersi, A.** 1, 2012, Media International Australia, Vol. 142, pp. 135-147.

593. *Good Jew, bad Jew ... good Muslim, bad Muslim: "managing" Europe's others.* **Topolski, A.** 12, 2018, Ethnic and Racial Relations, Vol. 41, pp. 2179-2196. Special Issue: Islamophobia and surveillance: genealogies of a global order.

594. *Islamophobia and the making of Latinos/as into terrorist threats.* **Romero, L. A. and Zarrugh, A.** 12, 2017, Global Trajectories and the Surveilance Order, Vol. 41, pp. 2235-2254.

595. *Propaganda and conflict: Evidence from the Rwandan genocide.* **Yanagizawa-Drott, D.** 4, 2014, The Quarterly Journal of Economics, Vol. 129, pp. 1947-1994.

596. **Brown, R.** *Prejudice: Its Social Psychology.* West Sussex : John Wiley and Sons, 2010. p. 368. 978-1405113069.

597. *Rationales for Employee Theft in Hospitality: Excuses, Excuses.* **Poulston, J.** 1, 2008, Journal of Hospitality and Tourism Management, Vol. 15, pp. 49-58.

598. *Employee theft as a reaction to underpayment inequity: the hidden cost of pay cuts.* **Greenberg, J.** 5, 1990, Journal of Applied Psychology, Vol. 75, pp. 561-568.

599. **Linden, S.** *A correlation analysis between employee theft and employee ownership (Doctoral Dissertation).* s.l. : University of Phoenix, 2009.

600. **Ruankqew, T.** *Ruankaew, Thanasak. The Impact of Job Satisfaction on Employee Theft among College Students.* Prescot Valley : Northcentral University, 2012.

601. **van Doel, R. M.** *Exploring Stewardship Characteristics in Professional,.* Marion : Indiana Wesleyan University, 2017.

602. *When more is not enough: Executive greed and its influence on shareholder wealth.* **Haynes, K. T., et al.** 2, 2014, Journal of Management, Vol. 43, pp. 555-584.

603. **Fiscal Policy Institute.** *Building Up New York, Tearing Down Job Quality: Taxpayer Impact of Worsening Employment Practices in New York City's Construction Indusry.* New York : Fiscal Policy Institute, 2007. p. 36.

604. **Chang, G. and Poo, A.-J.** *Disposable Domestics: Immigrant Women Workers in the Global Economy.* 2nd. Chicago : Haymarket Books, 2016. p. 272. https://www.amazon.com/Disposable-Domestics-Immigrant-Workers-Economy/dp/1608465284/ref=sr_1_1?s=books&ie=UTF8&qid=1544998762&sr=1-1&keywords=Disposable+Domestics. 978-1608465286.

605. *The erosion of minimum wage policy in Australia and labour's shrinking share of total income.* **McKenzie, M.** 2018, The Journal of Australian Political Economy, Vol. 81, pp. 52-77.

606. *Altered trust learning mechanisms among female adolescent victims of interpersonal violence.* **Lenow, J., Cisler, J. and Bush, K.** 1, 2018, Journal of Interpersonal Violenc, Vol. 33, pp. 159-179.

607. **Fiorentini, G. and Peltzman, S.** *The Economics of Organized Crime.* Cambridge : Cambridge University Press, 1997. p. 301. https://www.cambridge.org/us/academic/subjects/economics/microeconomics/economics-organised-crime?format=PB&isbn=9780521629553. 9780521629553.

608. **Francis, R. C., Holekamp, H. E. and Smale, L.** Dominance, demographics and reproductive success in female spotted hyenas: a long-term study. (pp 364-384). [book auth.] A. R. E. Sinclair and P. Arcese. *Serengeti II: Research, Management, and Conservation of an Ecosystem.* Chicago : University of Chicago Press, 1995, p. 673.

609. *Post-weaning maternal effects and the evolution of female dominance in the spotted hyena.* **Watts, H. E., et al.** 1665, 2009, Proceedings of the Royal Society B, Vol. 276, pp. 2291-2298.

610. *Behavioral processes and costs of co-existence in female spotted hyenas: a life-history perspective.* **Hoffer, H. and East, M. L.** 5, 2003, Evolutionary Ecology, Vol. 17, pp. 315-331.

611. *Rank-related maternal effects of androgens on behaviour in wild spotted hyaenas.* **Dloniak, S. M., French, J. A. and Holekamp, K. E.** 2006, Nature, Vol. 440, pp. 1190-1193.

612. *Sexual coercion and courtship by male western gorillas.* **Breuer, T., Robbins, A. M. and Robbins, M. M.** 1, 2016, Primates, Vol. 57, pp. 29-38.

613. *Sexually coercive male chimpanzees sire more offspring.* **Feldblum, J. T., et al.** 23, 2014, Current Biology, Vol. 24, pp. 2855-2860.

614. *Female–male relationships influence the form of female–female relationships in olive baboons, Papio anubis.* **Silk, J. B., et al.** 1, 2017, Animal Behaviour, Vol. 131, pp. 89-98.

615. *Coping with conflict: Testosterone and cortisol changes in men dealing with disagreement about values versus resources.* **Harinck, F., et al.** 4, 2018, Negotiation and Conflict Management Research, Vol. 11, pp. 265-277.

616. *Random walks on semantic networks can resemble optimal foraging.* **Abbott, J. T., Austerweil, J. L. and Griffiths, T. L.** 3, 2015, Psychological Review, Vol. 122, pp. 558-569.

617. *Human foraging behavior: A virtual reality investigation on area restricted search in.* **Kallf, C., Hills, T., & Wiener, J. M.** 32, 2010, Cognitive Science, Proceedings of the 32nd Annual Conference of the Cognitive Science Society, Vol. 32.

618. *Misinformation on the internet: Applying evaluation skills to online information.* **Fitzgerald, M. A.** 3, 1997, Emergency Librarian, Vol. 24, pp. 9-14.

619. *Reliability of information on the Internet: Some distinctions.* **Vedder, A. and Wachbroit, R.** 4, 2003, Ethics and Information Technology, Vol. 5, pp. 211-215.

620. *Ileal-lymphoid-nodular hyperplasia, non-specific colitis, and pervasive developmental disorder in children (retracted article).* **Wakefield, A. J., et al.** 9103, 1998, Lancet, Vol. 351, pp. 637-641.

621. *Controversial MMR and autism study retracted.* **McKee, M.** 2004, New Scientist.

622. **Deer, B.** Revealed: MMR research scandal. *The Times.* 2 22, 2004.

623. *Clear evidence of falsification of data should now close the door on this damaging vaccine scare.* **Godlee, F., Smith, J. and Marcovitch, H.** 2011, British Medical Journal, Vol. 342, p. c7452.

624. *Wakefield's article linking MMR vaccine and autism was fraudulent (erratum).* **British Medical Journal.** 2011, British Medical Journal, Vol. 342, p. d1678.

625. *Lancet retracts 12-year-old article linking autism to MMR vaccines.* **Eggerston, L.** 4, 2010, Cnadian Medical Association Journal, Vol. 182, pp. E199-200.

626. **Centers for Disease Control and Prevention.** Report shows 20-year US immunization program spares millions of children from diseases. *CDC Newsroom.* [Online] 4 24, 2014. https://www.cdc.gov/media/releases/2014/p0424-immunization-program.html.

627. *A planet orbiting the neutron star PSR1829–10.* **Bailes, M., Lyne, A. G. and Shemar, S. L.** 1991, Nature, Vol. 352, pp. 311-313.

628. **Lemonick, M. D.** When Scientists Screw Up. *Slate.* [Online] 10 15, 2012. https://slate.com/technology/2012/10/scientists-make-mistakes-how-astronomers-and-biologists-correct-the-record-when-theyve-screwed-up.html.

Made in the USA
Monee, IL
04 November 2026

67375825R00177